NOT
NECESSARILY
ROCKET
SCIENCE

NOT NECESSARILY ROCKET SCIENCE

A BEGINNER'S GUIDE TO LIFE IN THE SPACE AGE

KELLIE GERARDI

CORAL GABLES

Published by Mango Publishing Group, a division of Mango Media Inc.

Cover Image: © Jesse Dittmar, Kellie wearing a Final Frontier Design spacesuit
Cover Design, Layout & Design: Morgane Leoni

For permission requests, please contact the publisher at:
Mango Publishing Group
2850 S Douglas Road, 2nd Floor
Coral Gables, FL 33134 USA
info@mango.bz

For special orders, quantity sales, course adoptions and corporate sales, please
email the publisher at sales@mango.bz. For trade and wholesale sales, please
contact Ingram Publisher Services at customer.service@ingramcontent.com or
+1.800.509.4887.

Not Necessarily Rocket Science: A Beginner's Guide to Life in the Space Age

Library of Congress Cataloging-in-Publication number: 2020940933
ISBN: (print) 978-1-64250-410-1, (ebook) 978-1-64250-411-8
BISAC category code SCI098000, SCIENCE / Space Science

Printed in the United States of America

FOR DELTA V.

"There are no passengers
on Spaceship Earth.
We are all crew."

—Marshall McLuhan

CONTENTS

INTRODUCTION

When I was young, I often wondered how people in the Renaissance regarded their own era. Did the general public clock the rise of polymaths like da Vinci or Michelangelo as historic humans? Could they feel the tug of modernity as the medieval world melted into the past? It's doubtful that anyone extrapolated a global cultural awakening from the freshly painted ceiling of the Sistine Chapel. It's even more unlikely that anyone in the Stone Age paused pummeling their dinner with a club to think about the profound impact of that percussion on the trajectory of the entire human species. Denizens of the Enlightenment were not explicitly told that they were entering an Age of Reason and folks in the Industrial Revolution buzzed about their days unaware that their highly productive lives coincided with one of the major innovational turning points in all of human history.

Of course, each of these eras was named after the fact, labels of approximation rather than precision, a set of belated bookends for more modern humans to organize history and figure out how we surpassed our humble roots to become the mightiest species on Earth. But might the velocity have been different if everyone, not only the history book celebrities, had considered their own individual role in that journey and the opportunity in front of them? What might societies have done differently if individuals were privy to the trajectory of the entire species, or better yet, equipped from birth with a handbook about their generation's unique stop on the human journey? Welcome to the world, tiny human, and listen up!

What a time to be alive then, to recognize our own era within history. If no one has shared the good news yet, please let me be the first to welcome you to a full-blown Space Age! The human species is on an astronomical trajectory and it would be an honor to take you on a brief tour of our very own moment in history as it's happening. This is our chance to find out how much momentum might be gained if *everyone*—not only the rocket scientists—operates with the same level of awareness around this incredible window of opportunity. Consider this a beginner's guide to life in the Space Age.

Art was only one manifestation of a new way of thinking in the Renaissance. Cultural innovation was equally apparent across the vastly different disciplines of medicine, technology, religion, politics, philosophy, science, and even warfare. Similarly, engineering feats represent one small slice of the Space Age. Future historians will widen the lens on a broader cultural movement that saw twenty-first-century humans contemplating our next giant leap as a species, marking the beginning of our transition from the Earth to the stars.

It should be obvious then that the future of our species doesn't rest solely on the shoulders of rocket scientists; like any turning point in history, humanity's next giant leap will require the contributions of artists, engineers, and everyone in between. For the first time in more than 4.5 billion years, life on Earth has the ability to venture beyond this planet—the potential to become interplanetary and secure a long-term survival in the cosmos. Thanks to the rise of the commercial spaceflight industry, routine space travel has become a dream within anyone's reach. My own career is a testament to that democratized access: as a non-engineer I went from dreaming about the promise of space exploration to contributing to it, and eventually to training for it myself.

I'm sharing my own experiences and reflections in the hopes that they might spark your own passion for exploration and discovery. At the very least, I hope you find yourself in the conversation and recognize your rightful place in the Space Age. Human spaceflight is about more than simply satisfying curiosity and inspiring dreams; it's also about ensuring the survival of a species whose home planet has an expiration date. At the more generous estimates, we're looking at a few billion years, when our sun ceases to provide its nurturing energy for life on Earth. Or perhaps it's a few million years, when another rock along the size of a dinosaur-level extinction makes impact. Or maybe it's a whole lot sooner, a Russian roulette of global pandemic, massive ecological collapse, or nuclear obliteration with just the push of a button.

But the human species is nothing if not resilient. For 200,000 years we've moved forward together, a species both mission-driven and coordinated. From the moment we stood upright, we've been forced into a fighting stance against nature, disease, predators, and perhaps most viciously, ourselves. For millennia we've harnessed our collective force to defy the odds and propel ourselves into the future. Our scrappiness and grit saved us from mass extinctions and an Ice Age that nearly extinguished our journey before it began. Our cleverness earned us millennia of cultural and scientific advancement, and most recently, a very promising start to this Space Age. Together we've made glory worth pursuing, new frontiers worth exploring, and survival worth fighting for. We've put up one hell of a fight.

And now, through luck of birth, you and I find ourselves at the starting line of the final frontier. We're holding the baton of survival that has passed through the hands of 10,000 generations of humans before us. Too many times throughout history has that baton almost dropped and the spark of life been extinguished. But at each baton fumble, another hand swooped in to secure it. Sometimes that hand belonged to an engineer, inventing tools to advance us or medical breakthroughs to heal us; other times the hand belonged to an artist, creating the language to connect us or the culture to civilize us. Encore! The survival of our species has always depended on a diversity of talent and contributions, and damned if we're going to let the baton drop on our watch.

Each one of us has a role to play in humanity's next chapter. You don't need to be an astronaut to feel goosebumps during the countdown of a rocket launch or a surge of adrenaline watching humans take flight; all of us are wired to appreciate the profundity in the sights and the sounds of the final frontier. Our subconscious recognizes that to launch something off of this planet is a uniquely human act of resilience, a primal survival instinct that has carried the species all the way to the twenty-first century.

But before we contemplate our place in the Space Age and gear up for humanity's next giant leap, we should revisit the humble beginnings of the greatest marathon in the Milky Way. It's a story you've likely heard before, certainly bits and pieces. I want to share a brief history of you and me, a tale of our shared origin and destiny as a spacefaring species. And if you'll indulge me, I'd like to start at the very beginning.

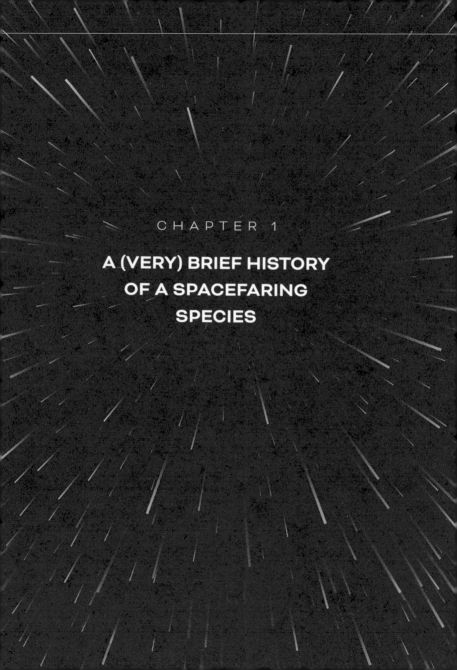

CHAPTER 1

A (VERY) BRIEF HISTORY OF A SPACEFARING SPECIES

AROUND 4.5 BILLION YEARS AGO, A MASSIVE STAR EXPLODED. NOT ORDINARILY A BIG DEAL— happens all the time across the universe. But this particular doomed star lived near the Milky Way galaxy, and its explosion rocked to life our cold, dusty, and previously empty corner of the universe.

Details are hazy this far back, but our best guess on how things played out goes something like this: The shock wave sent dust particles and gas remnants swirling. Gravity yanked on that cloud of stardust, condensing it and whipping it around until its particles spun into a white-hot disc. The blazing center of that disc would become our sun, and the fiery clumps flinging off the perimeter would cool off over the course of a couple hundred million years, taking the eventual form of the planets and solar system we know today.

From the beginning, Earth was a special planet. A respectable distance from that new sun, early temperatures were not too hot, not too cold, and provided just the right conditions for liquid water. Earth was a true "Goldilocks" planet where the building blocks of life were set up to flourish and evolve. The exact mechanics of that evolution are also hazy, but some evidence suggests all life on Earth might be traced back to a single ancestral cell or population of cells, a biological supernova that ignited the spark of life some 3.5 billion years ago. Not bad for a random cloud of reheated stardust.

From that first spark, Earth was ablaze with life. A few billion years passed as microbial kingdoms gave way to plant and animal kingdoms. And by 3.5 million years ago, the early signs of human life had already emerged: a group of *Australopithecus afarensis* (Latin for "southern apes from Afar") stood upright and strolled across the plains of Laetoli, Tanzania, their footprints preserved in the volcanic ash. One small step for early hominins and one giant leap for humankind.

Taller and more capable, a long line of early human species carried on the bipedal march of survival, acquiring new capabilities along the way. For instance, a clever *Homo habilis* ("skilled human") can be credited with kicking off the Stone Age just over two million years ago, striking stones together with precision and inventing tools in the process. Then *Homo erectus* ("upright human") harnessed the power of fire for safety, warmth, and cooking. And *Homo neanderthalensis* (Neanderthals), our closest extinct human relative, demonstrated not only a sophisticated use of tools and control of fire but also a touching sense of sentimentalism, deliberately burying and commemorating their dead.

But none of these starter species proved a match for *Homo sapiens* ("wise human"), who hit the African scene some 200,000 years ago and spread out from there, capitalizing on generations of ancestral knowledge to adapt to their environment and rise to meet its many challenges. As each of those early species hit dead-ends and dropped their batons, *Homo sapiens* tightened their grip. Big brains and a peak perch atop the food chain afforded modern humans the luxury of introspection. At some point, likely huddled around a well-tended fire, our early ancestors attempted to make sense of this mysterious world and our place within it. Early interpretations were imaginative and sparked existential questions about our origin and destiny.

In modern times, we recognize the cyclical nature of our journey: we came from the stars, and if we are to continue our long streak of survival, back to the stars we must one day return. But we still had a long way to go.

Understandably, early humans both feared and worshipped the night sky. Our brains were still growing but our egos were fully developed; we believed we were the center of the universe and that cosmic events were directly related to our performance. Picture the shock of a solar eclipse or the spectacle of a meteor shower streaking through the darkness and you can sympathize with the assumption that gods themselves must be pulling the strings, using the sky as a canvas to convey praise or rain down punishment.

Ancient cultures painstakingly recorded mysterious celestial phenomena, and in this way, astronomy became our first and original science. The Babylonians were a particularly scholarly bunch, taking an empirical approach to astronomy and carving their observations on clay tablets over a period of centuries. These "Astronomical Diaries" formed the first functional theory of planets and served as a nascent national security program for the Kingdom of Babylonia, helping them predict, interpret, and plan around celestial events.

Records, humans discovered, were supremely useful. A millennium later, the data would fall into another pair of ambitious hands. The ancient Greeks reasoned that there must be an underlying cosmological framework that tied everything together, and an astronomical almanac from the Babylonians made for perfect reference material. Mythology inspired philosophy, and before long philosophers began thinking scientifically.

The sequence went something like this: some 2,500 years ago, Greek philosopher and mathematician Pythagoras suggested a spherical Earth, rather than a flat one. The famous Plato would echo the same belief a century later, greatly popularizing the theory. But it was Plato's star pupil, Aristotle, who became the first to offer actual evidence, citing horizon lines and eclipse shadows. A century after that, Eratosthenes would take those observations further, managing to calculate the circumference of this round rock we call home using only sunlight and a stick. For those keeping score, we're still the center of the universe, but we're round now. So far, so good.

Fast-forward nearly two thousand years though, and early signs of controversy emerge in the form of Nicolaus Copernicus, who surprised sixteenth century Renaissance culture with an updated model of the universe when he published *On the Revolutions of the Heavenly Spheres*. His heliocentric theory broke with centuries of traditional belief in placing the sun—not the Earth— at the center of the universe. This was potentially dangerous territory given the Catholic Church's teachings to the contrary, but because it was a dense mathematical argument limited almost entirely to scholarly circles, the overall reaction was one of mild derision rather than harsh condemnation.

That leniency would expire a century later when Galileo Galilei turned the newly invented telescope toward the night sky for the first time. Armed with observational evidence, he thrust the controversial Copernican theory into the seventeenth-century spotlight, carrying the conversation far beyond academic circles.

His conclusions were visibly and defiantly against those of the Church, who now felt compelled to respond. Seventy-three years after its release, Copernicus' book was withdrawn from circulation and the heliocentric theory was declared to be "foolish and absurd, philosophically false and formally heretical." Ignoring warnings, an indignant Galileo then defied the pope's orders and published a rebuttal in *A Dialogue on the Two Chief World Systems*, which surged in popularity and further provoked the Church. Galileo and the pope sat down to calmly discuss their divergent points of view and agreed to disagree, parting with a handshake.

Just making sure you're still with me. What actually happened was this: the book was immediately banned, and Galileo was hauled in by the Roman Inquisition and made to stand trial. Disgraced threatened with torture, he recanted his heresy in exchange for leniency, swearing his commitment to the Church and vowing never again to to "say or assert, verbally or in writing, anything that might furnish occasion for similar suspicion." The rest of his life was spent under house arrest, until his death in January of 1642. An unfortunate end for someone who contributed so much to science, but his stubbornness helped pave the way for another scientific luminary to pick up where he left off. On Christmas Day of that same year, *Homo sapiens* welcomed the newest member of the species, a baby boy named Isaac Newton.

Newton's "ah ha!" moment had modest beginnings. An apple falling from a tree inspired his theory of gravitation, a concept he would build upon in *Mathematical Principles of Natural Philosophy* to lay out the universal laws of motion. But if Galileo's teachings were burns to the Church, Newton's were a salve. A man of faith, he cautioned that gravity may explain the motions of the planets, but it says nothing of who set them in motion in the first place. "God," he insisted, "governs all things and knows all that is or can be done." A uniquely *Homo sapien* display of compromise.

Newton's contributions earned him recognition as one of the greatest scientists of all time, and his explanations of motion and gravitation would represent our most comprehensive scientific understanding for the next two centuries, until Albert Einstein shared the theory of general relativity, which introduced radical concepts like the curvature of space and time.

By the 1900s, *Homo sapiens* had a toolbox that surpassed the wildest dreams of the now distant *Homo habilis*. Modern humans were suddenly capable of solving some major cosmological problems, but it was far beyond the power of mathematicians alone to motivate society toward unlocking these possibilities. Instead we had to look to the storytellers, because far before the science was settled, the science fiction was sparking imagination.

🚀 🚀 🚀

My personal dream of spaceflight didn't start in a science classroom or at the rocket company where I eventually worked. Like so many other readers, it started as a child under the covers with a flashlight, up way past my bedtime, engrossed in far-out adventures and wondering what it would be like to explore the cosmos myself.

I devoured the twentieth-century greats: I cut my teeth on Isaac Asimov's *I, Robot* and Arthur C. Clarke's *Space Odyssey*, and at one point felt certain I would never discover a novel as exciting as Robert Heinlein's *Stranger in a Strange Land*. Their combined literary genius pioneered the genre of hard science fiction, but like all exceptional humans, these authors stood on the shoulders of those who came before them. A common ancestor, if you will.

In the 1800s, French author Jules Verne was pushing the boundaries of travel writing. His magazine adventure tales were meticulously researched and his penchant for accuracy inspired an idea for a new literary work: a novel of science, in which he could convey an exciting blend of scientific detail and imagination. What followed were fantastic adventures like *Journey to the Center of the Earth*, *Twenty Thousand Leagues Under the Sea*, and *Around the World in Eighty Days*. A generation later, H. G. Wells would similarly captivate imaginations around the world with far-out tales like *The Time*

Machine and *The War of the Worlds*. Together, their bodies of work exerted a profound influence on the emerging genre of science fiction, and soon after, the science itself.

Scientific innovation has long been driven by the imaginations of authors like Verne and Wells. Robots, holograms, and spaceships appeared in their novels long before their physical inventions. But the power of science fiction writers isn't their ability to predict the future, it's their ability to inspire it—when an adventure is so vivid and compelling that it takes root in a reader's imagination until the reader eventually wonders, "what might it take to bring this to life?"

That was the spark of possibility that blazed in the mind of a young American boy named Robert Goddard, who after finishing *The War of the Worlds*, dreamed of one day building a spacecraft of his own. His doctorate studies in physics afforded him the use of a small laboratory at Clark University where he could carry out his rocket engine experiments, including some early explosions which caused significant alarm. Goddard was certain that it was possible to make a rocket powerful enough to escape Earth's gravity, and perhaps even fly all the way to the Moon. And although it would take years for him to prove the skeptics wrong, and for the government to eventually begin sponsoring his experiments, Goddard committed himself to tinkering with engine refinement. He reasoned that powder rockets were holding him back and costing efficiency, and by March of 1926 his conviction paid off: the world's first liquid-fueled rocket took flight from his Aunt Effie's farm in Auburn, Massachusetts. The flight lasted less than three seconds and reached a peak of altitude of just forty-one feet, but it was a critical demonstration of the potential for liquid propellants. Just like that, modern rocketry was born, not with a bang, but with a nosedive into a cabbage field.

Scientific progress isn't always linear; sometimes it's simultaneous. Because on the other side of the globe, at the very same time that Goddard was first flipping through the pages of *The War of the Worlds*, a Russian man named Konstantin Tsiolkovsky was nurturing a similar spark of possibility that had taken root in his mind as a child, when he first read Verne's *From Earth to the Moon* and puzzled over what it might take to make real that fantastic story of human spaceflight.

Tsiolkovsky had lamented that the space cannon of Verne's imagination would require an impossibly long barrel to work in real life but wondered about other possible methods of launching a human to the Moon. Over the course of hundreds of publications, Tsiolkovsky gifted humanity with a comprehensive theory of spaceflight and designs for a multistage rocket that would get the job done. His 1903 "Exploration of Outer Space by Means of Rocket Devices" debuted the Tsiolkovsky equation and provided proof that a multistage rocket running on liquid hydrogen and liquid oxygen could achieve the speed necessary for an orbit around the Earth.

Having solved the practicalities of a launch from Earth, Tsiolkovsky turned his attention to the environment of space, detailing designs of spaceships, space stations, colonies, and life-support systems that could protect us from such an inhospitable vacuum. His motivation was anchored in the belief that the Earth was only the beginning of humanity's journey. There were great benefits to reap from the rest of the universe and great risk in failing to do so. "We have said a great deal about the advantages of migration into space," he warned. "But not all can be said or even imagined."

Without space travel, he cautioned, humans would be sitting ducks for numerous and terrible dangers that could "wipe off the face of the Earth all traces of man and his buildings." He warned of devastating infectious diseases, rising temperatures, resource depletion, and even asteroid impact. And if we managed to somehow survive all of that? Still doomed, because our sun itself is a dying luminary. Tsiolkovsky was far ahead of his time: he realized that life on Earth has an expiration date and investing in space exploration was our key to long-term survival. But with war brewing back down on Earth, there were much more near-term threats to consider, and people were starting to realize that space technology could be used for a whole lot more than exploration.

While this is an extremely abridged tour through our spacefaring history, a moment must be spent on the topic of World War II, which earned the distinction of being the most destructive conflict in all of human history. An impossible-to-fathom 75 million people lost their lives, close to 3 percent of the entire world's population at the time. Civilian deaths accounted for more than double the deaths of military personnel, largely due to genocide, mass bombings, and the second-order effects of disease and starvation. This was a dark and deadly episode for the *Homo sapiens* species; along with many other consequential outcomes, World War II showed us our own fragility and how technology can be used for good, evil, and the moral ambiguity in between. Above all, it served as a sober reminder that no matter how hard-won our continued existence over the course of hundreds of thousands of years, there could come a day when we knock the baton of survival out of our own hands, with no one left to swoop in and catch it before it rolls out of reach. No encore: a final curtain call on what was once an extremely promising species.

But let's jump back to war-torn Europe, where rocketry was emerging as an extremely valuable skill. Just as Verne and Wells inspired Tsiolkovsky and Goddard, so too did their work influence others, including a talented German rocket scientist named Wernher von Braun. Like Tsiolkovsky and Goddard, von Braun became obsessed with the idea of spaceflight as a young man and turned his ambition toward a flight to the Moon. It was toward that goal that he pursued an engineering degree, during which time the German Army took notice of his talent and made the generous offer to finance his studies if he would agree to work on their nascent liquid-fueled rocket program. The sponsorship could have been a win-win, except shortly after von Braun agreed to the arrangement, Adolf Hitler was elected chancellor of Germany.

As Nazi Germany rose in power, von Braun rose in rank, and as the technical director of a new rocket center dedicated to designing and developing the world's first long-range guided ballistic missile, he accepted a formal invitation to join the SS as an officer. Government funding and support for his program grew, and in 1942 that missile finally took flight. Unlike Goddard's early experiment in the cabbage field, von Braun's liquid-fueled rocket engine was fully operational, and the world was introduced to an incredibly powerful weapon: the V-2.

Production of the weapon intensified, relying largely on forced laborers and concentration camp prisoners to produce the thousands of V-2s that the German Wehrmacht would launch against Allied targets from London to Antwerp. The missile's performance envelope broadened, and in 1944 a test launch reached an impressive altitude of 176km before crashing back to Earth. While it didn't reach orbit, the weapon of war became the first ever human-made object to cross briefly the boundary of space.

By this point, von Braun's ambition for achieving his youthful dream of spaceflight was likely dampened by the grind of war and Germany's deterioration. Nazi Germany would be defeated the following year, and as the country collapsed, there was an Allied scramble for the powerful technology. Von Braun organized a surrender to American forces and (along with 1,600 other elite German scientists and engineers) he was brought to the United States, where he would ultimately find the opportunity build his spaceships in a country where his Nazi background was not widely known.

In spite of the violence, *Homo sapiens* had still managed to tee up a Space Age. We dreamed of the stars from our earliest days and by the twentieth century we had finally invented the tools and devices to reach them. So it's quite the cosmic irony that we, the mightiest species on planet Earth, unlockers of space travel, would not earn the distinction of being the first species to break the chains of gravity.

No, that great galactic honor would belong to *Drosophila melanogaster*: the common fruit fly.

With V-2 technology secured, the Americans were confident in their ability to cross the boundary of space and perhaps even send a human one day. But what would happen once they got there? The environment of space was filled with unknown challenges and threats, and one of the more pressing questions was the effect of radiation exposure at that altitude.

The post-flight analysis of seeds and plants could only tell us so much. To truly begin to unpack the mysterious effects of space travel, we would need to send living passengers, ideally ones who contained some genetic similarities to human beings. And that is how, after enjoying forty million years of a relatively uneventful existence on Earth, the humble fruit fly became the first species to go boldly where no other species had gone before.

From the area now known as White Sands Missile Range in New Mexico, scientists loaded the tiny passengers into a capsule atop a V-2 rocket and watched them soar nearly seventy miles into the sky. Three minutes later, the capsule ejected and deployed its parachute before gently returning back down to Earth, cementing the fruit fly as the first species to not only visit space but to safely return. Unfortunately, the same could not be said for a number of species that would unwillingly follow in their footsteps.

Albert II, for example, became the first primate to reach space in 1949, but died on impact after a parachute failure botched his return. And as his name implies, there was indeed an Albert I, a predecessor who never made it off the ground due to a catastrophic launch failure. Alberts III and IV didn't fare any better, and the following year, the ill-fated Albert V was swapped out for a mouse whose rocket disintegrated after yet another parachute failure. The Russians were equally committed to sending living beings to space and found early success with two dogs, Tsygan and Dezik, who became the first higher organisms (no offense to the fruit fly) to be safely recovered from spaceflight. Breaching the boundary of space was exciting, but the Americans and the Soviets both knew that achieving orbit around the Earth would be the real prize.

With World War II in the rearview mirror and American sentiment toward Germans beginning to improve, von Braun found himself with a soapbox to share his ambitious space dreams. To an enthralled American public, he spoke of moon landings and space stations and even human missions to Mars. The first step would be to leverage his extensive missile experience to develop a rocket capable of entering into orbit around the Earth, but the US government wasn't immediately sold on the investment. It would take a few more years of short, suborbital test flights and directional indecisiveness until a swift kick in the butt propelled America's space program to life in 1957. That convincing kick came in the form of Sputnik 1, the first artificial satellite to orbit the Earth, and an ominous demonstration of space capability from the Soviet Union.

The US would establish by law a National Aeronautics and Space Administration nine months later, and soon prove that rapid advancement in space all comes down to the right motivation: the Space Race had begun.

Sputnik was a whole new ball game. While the American military was aware of the Soviet Union's maturing capabilities, the American public was blindsided. Throughout its twenty-one days of battery life, the beach ball-sized satellite transmitted an ominous and continuous "beep-beep-beep" back to Earth, which could be picked up by anyone with a shortwave radio receiver. The confidence in America's status as a technological superpower was wavering and it gave way to new fears about the growing capabilities of the Soviet Union. If they could put a satellite in space to circle the Earth, what else could they put up there? As it would turn out, the near-term answer was a stray dog named Laika.

From the beginning, Laika's flight was intended as a one-way mission. The ability to reach orbit had been achieved, but the technology to safely de-orbit had not yet been invented when she was strapped into her flight harness on Sputnik 2. Equipped with seven days' worth of food and oxygen, the best-case scenario would be for Laika to experience a week of orbital flight before submitting painlessly to oxygen deprivation. Unfortunately, she only made it a few hours. Soon after launch, a damaged heat shield spiked the temperature in the capsule, which quickly exceeded ninety degrees. By her fourth orbit, Laika met her fate as a sacrifice to science.

Meanwhile, Project Mercury was gaining momentum in the United States, where scientists were hard at work catching up to the Soviets. Americans had matched Sputnik 1 and 2 with Explorer 1, but the goal was now human spaceflight. More testing was needed, and the Holloman Aerospace Medical Center knew just the chimp for the job. Ham, named after his facility, would be the first hominid in space.

Ham was no mere passenger—he had an active role to play during his flight. In the months before his flight, punishment and reward training in the form of mild shocks and banana treats led Ham to master a number of timed tasks, like learning to push a lever at the prompting of a light. By the time he got to space, he was pushing levers at nearly the same speed as his Earth training, demonstrating that humans should be perfectly capable of performing tasks in space. After sixteen minutes in flight, Ham splashed down safely off the coast of Cape Canaveral, Florida, and enjoyed another twenty-five years of celebrity.

The Space Race was in full swing by this point, and after a multi-year procession of fruit flies, monkeys, mice, and dogs, it was finally time for humans to leave the cradle of Earth's atmosphere. That first honor would belong to Soviet cosmonaut Yuri Gagarin, whose *Vostok 1* capsule completed one orbit of the Earth on April 12, 1961. After 200,000 years of exploring our home planet, *Homo sapiens* was officially a spacefaring species, and the years that followed were an exciting blur of achievement and competition. A month after the Soviets launched Yuri Gagarin, American test pilot Alan Shepard completed a suborbital flight aboard *Freedom 7*, and a brand-new profession made its way into the American lexicon: astronaut.

The fact that the United States hadn't yet managed to put a human into full orbit did nothing to dissuade confidence. Less than three weeks after Shepard's flight, in a special joint session of Congress, President John F. Kennedy set America's sights even higher and requested funding for the most ambitious goal the world had ever seen: before the decade was over, America would land a man on the Moon and return him safely to Earth. Kennedy's vision was prescient, promising, "No single space project in this period will be more impressive to mankind, or more important in the long-range exploration of space; and none will be so difficult or expensive to accomplish." Sadly, he wouldn't live to see this incredible achievement realized. Two years later in 1963, President John F. Kennedy would be assassinated, and Lyndon B. Johnson would be sworn in as the leader responsible for carrying forward this bold vision and guiding the species on a moonshot.

Before America reached for the Moon, there were still a number of details to be worked out in low Earth orbit. The Soviets launched *Vostok 2* with Gherman Titov, who not only became the first human to spend more than a day in space, but also the first to vomit in space. Despite the space sickness, the longer flight was good news for America's lunar plans: his record-breaking seventeen orbits proved that humans were indeed capable of living and working in space. The US followed closely behind, eventually achieving an orbital flight of their own with *Friendship 7*, which carried Mercury astronaut John Glenn around the Earth. It was a fine achievement, but the Moon was still far away, and America had to work fast to make good on its promise.

CHAPTER 2

MOONSHOTS

BY 1967, THE APOLLO PROGRAM WAS HUMMING, AND GUS GRISSOM, ED WHITE, AND Roger Chaffee had their crew assignment: they were to carry out the first crewed test mission of the Command Module. The CM was a spacecraft more complex than ever before, and hundreds of engineering design modifications were required in the year leading up to the test flight. As critical design issues continued to surface, the *Apollo 1* crew remained closely involved, frequently making suggestions and voicing concerns of their own. It was a relief when the vehicle finally passed its altitude chamber test—a sprint to the finish line, but now the only thing standing in between *Apollo 1* and its crewed test flight was one final dress rehearsal on the pad to confirm the spacecraft could function on its own power.

Donning full pressure suits, Grissom, Chaffee, and White strapped in and the hatches were sealed, while pure oxygen pumped through the capsule. The spacecraft had a number of issues, including microphone static that was impeding communications. The simulated countdown was put on hold while attempts were made to repair the issue. "How are we going to get to the Moon if we can't talk between two or three buildings?" Grissom snarked. The crew used the time to run through their checklist one last time, but seconds later, shouts rang out from the capsule: "Hey! ...got a fire in the cockpit!" This was followed a few seconds later by a patchy transmission many listeners would interpret as "We've got a bad fire—Let's get out... We're burning up!" There was an anguished cry before communications dropped altogether.

By the time pad workers forced open the hatch, it was too late: all three astronauts had suffered third-degree burns and succumbed to asphyxia from the inhalation of toxic gases in the cabin. The crew had attempted to open the hatch but couldn't manage it against the immense interior pressure. The devastation was immediate and the fallout swift. A review board would identify a number of factors whose combination was determined to have caused the fire, including vulnerable wiring, the pure oxygen cabin environment, an excess of combustible materials located near ignition sources, a hatch cover that became near impossible to remove in a high-pressure environment, and inadequate emergency preparedness. The hard lessons led to procedural overhauls and increased scrutiny at NASA. It was a tragedy that drove home the grave risks inherent with spaceflight and the great cost of opening new frontiers. But the Apollo program persevered, and with only three years left in the decade, quite a bit of work remained.

🚀 🚀 🚀

Thankfully, there were plenty of talented hands on deck. By this point, the Apollo efforts employed nearly 400,000 women and men across the US, along with the support of more than 20,000 companies and universities. If it takes a village to raise a child, it takes a nation to launch a moonshot. Beyond the more visible astronauts, mission controllers, and engineers, the heartbeat of Apollo would be sustained by the efforts of folks working largely behind the scenes. Films like *Hidden Figures* have helped shine a light on the invisible labor that had previously been written out of history, such as the largely female "human computer" workforce, the most renowned of whom were Black women. Beyond mathematics and programming, America's moonshot was further propelled by an ecosystem of financial analysts, public affairs professionals, physicians, administrative support staff, and so many more essential contributors. Not to mention the countless journalists, broadcasters, authors, editors, and photographers who sustained above-the-fold coverage and ensured that Apollo remained an American dinner table topic over the course of an entire decade.

The Moon was within our reach and a series of un-crewed preparatory missions moved us even closer: a launch of the mighty *Saturn V* rocket, a test of the Lunar Module in Earth orbit, and an incomplete but satisfactory dry run of the propulsive maneuver that would get the whole package to the Moon. By the fall of 1968, the *Saturn V* was human-rated and ready for crewed test flights. The tests were incremental but thrilling: the crew of *Apollo 7* completed an eleven-day test flight and earned an Emmy Award for their live broadcast from Earth's orbit; the *Apollo 8* crew became the first humans to enter lunar orbit and captured the world's attention on Christmas Eve with pictures of the lunar surface; the crew of *Apollo 9* practiced rendezvous and docking, along with a test of the spacesuit outside the Lunar Module; and the *Apollo 10* crew carried out the final dry run, guiding the Lunar Module to a tantalizing fifteen kilometers away from the lunar surface.

And then, on July 20, 1969, America's moonshot goal was finally realized with *Apollo 11*. Astronaut Michael Collins orbited the Moon in the Command Module while crewmates Neil Armstrong and Buzz Aldrin stepped down from their Lunar Module onto the surface, some quarter of a million miles away from their home planet. As the first human to ever step foot on another celestial body, Neil Armstrong delivered the famous line that summed up the profound impact on our species: "That's one small step for a man; one giant leap for mankind."

An estimated 600 million people—nearly 15 percent of the world's population—watched those steps in more or less real time, tuning in from all around the globe to watch humanity's first foray in the cosmos. For a few incredible hours, astronauts bounced around the surface of another world, collected surface samples, and set up scientific instruments. In addition to an American flag and a mission patch paying tribute to the crew of Apollo 1, America's first moonwalkers left behind a plaque to mark the momentous occasion, which included their own signatures along with that of newly elected US president Richard Nixon:

HERE MEN FROM THE PLANET EARTH
FIRST SET FOOT UPON THE MOON
JULY 1969, A.D.
WE CAME IN PEACE FOR ALL MANKIND

Armstrong referred to the flight as the "beginning of a new age," and for three more years the Apollo program flourished. Plans were made for lunar landing missions all the way through *Apollo 20*, but difficulties and cutbacks loomed. *Apollo 12* made it to the Moon without a hitch, but two days into *Apollo 13*'s mission an oxygen tank exploded, forcing the crew to use the Lunar

Module as a sort of lifeboat back to Earth. Oxygen tank redesigns temporarily paused the program and the momentum was never quite recaptured.

The dissonance between the activities on Earth and in space was noticeable. As the nation refocused on the Vietnam War and civil rights, public support for space funding began to wane. Moonshots had a price tag, people realized, and that price tag was about twenty-five billion dollars that many folks felt could be put to good use fixing problems right here on Earth. By 1970, with a shifting national agenda and no further Cold War competition on the horizon, the final three Apollo missions were scrubbed, leaving only four more opportunities for astronauts to explore our planet's moon. *Apollo 14, 15,* and *16* made the most of it, extending time on the lunar surface, collecting more samples, and making use of a Lunar Rover to cover more ground, all teeing up *Apollo 17* to return for one final visit.

The last Apollo mission included an unlikely crew member in Harrison Schmitt, the first professional scientist to go to space. Formally trained as a geologist, NASA selected Schmitt in their first ever class of "scientist-astronauts," instead of the usual test pilots. Realizing this as our last shot at learning more about the mysterious lunar surface, America finally sent a scientist to the Moon, and his selection set an important precedent for academics in the astronaut corps.

As the Apollo program faded in the 1970s, Nixon rejected proposals for sustained human presence on the moon, and his rationale foreshadowed the limitations of relying solely on government funding for space exploration. "We must build on the successes of the past, always reaching out for new achievements," he explained, "but we must also recognize that many critical problems here on this planet make high priority demands on our attention and our resources."

The mysteries of the Moon whetted our appetite for the rest of the solar system, but human space exploration beyond low Earth orbit was extremely expensive. Without the humans, though, we could continue our reach for those newer, further achievements— and at a fraction of the cost.

By the late 1970s, Apollo budgets were a thing of the past, but space was still front and center for the American public. *Star Wars* was breaking box office records and real-life robotic spaceflight was picking up steam. Having already sent a probe past Mercury and Venus, a group of NASA scientists now had their eyes set on the outer planets. A rare alignment of Jupiter, Saturn, Uranus, and Neptune was approaching, and if timed just right, NASA could send a spacecraft on a tour of multiple planets for the price of one, using a fuel-saving "gravity assist" technique to swing the probe from one planet's orbit to the next. The Voyager program would carry humanity further than ever before, and the program's twin spacecraft would become our eyes and ears in deep space.

The primary objectives of *Voyager 1* and *Voyager 2* were close flybys of Jupiter and its moon Io along with Saturn and its moon Titan, but the chosen trajectories preserved the feasibility for

Voyager 2 to continue on to Uranus and Neptune if budgets allowed. The knowledge gained from early observations of Jupiter was explosive: its large moon Io revealed the presence of giant, active volcanoes shooting plumes more than 190 miles above its surface. And Jupiter's "Great Red Spot," which astronomers had been observing for more than a century through the lens of a telescope, was revealed to be a gigantic and slow-moving storm, twice as wide as the Earth itself. During a flyby of Saturn, three new moons were discovered, along with the first close-up observations of the planet's famed rings. The probes had already returned enough data to rewrite science textbooks, and the opportunity to continue on was too tantalizing to pass up. *Voyager 2*'s mission was extended to Uranus and Neptune, providing our first close-up views of the mysterious gas giants.

The avalanche of scientific discoveries from the *Voyager 1* and *2* changed the course of planetary science, but it was art that really elevated the profile of the missions and captured the imagination of the public. Like our Neanderthal ancestors, *Homo sapiens* are a sentimental bunch. It wasn't lost on the science team that at some point after flying past their mission targets, *Voyager 1* and *2* would become the first human-made objects to leave our solar system, a pair of interstellar emissaries from planet Earth. We had no idea what the Voyagers might discover as they hurtled through deep space, exploring the cosmos far beyond our ability to communicate with them, but the journey presented a once-in-a-lifetime opportunity for first contact. With just nine months to go until launch, famed astronomer Carl Sagan convinced NASA to include a greeting for any extraterrestrial civilizations that might encounter these two plucky probes in the future.

Where to begin with a message of that magnitude? Sagan immediately assembled a team for the impossible task of selecting the sounds and images intended to represent all of life and culture on planet Earth. The format would be a twelve-inch gold-plated phonograph affixed to each of the probes, and like any mixtape, space was limited. Narrowing the story of humanity down to ninety minutes of music and 118 images to be etched onto the disc was an agonizing level of responsibility. "I remember sitting around the kitchen table making these huge decisions about what to put on and what to leave off," collaborator Ann Druyan recalled to NASA. "We couldn't help but appreciate the enormous responsibility to create a cultural Noah's Ark with a shelf life of hundreds of millions of years." Druyan and Sagan would also begin a romance during the whirlwind project, and as a sentimental addition, the pair managed to squeeze in a short soundbite featuring the EEG brainwaves of a woman in love.

The record's cover art doubled as an instruction manual for a far-future recipient interested in unlocking the sights and sounds of planet Earth, and the images themselves ranged from educational to artistic. The slideshow kicked off with diagrams and reference guides to our solar system, DNA structure, mathematics, cells, and human anatomy, followed by a diverse assortment of landscapes, animals, children, families, social activities, buildings, cities, and even rocket launches. It was a brief but impressive visual tour through the evolution of humanity.

On the musical side, Bach, Beethoven, Stravinsky, and Mozart made the cut, as did snippets of traditional arrangements of folk music from cultures around the world. Selecting music was difficult and untangling the first literal application of "universal rights" with record labels was also a challenge. After much deliberation, Chuck Berry's "Johnny B. Goode" and Blind Willie Johnson's "Dark was the Night" made the contemporary list. Beyond music, the records contained audio essays of life on Earth, including the sounds of animals, weather, footsteps, machinery, laughter, a mother's kiss, and simple greetings in fifty-five languages.

Those simple greetings didn't fall together so simply, though. Sagan first aimed to record United Nations delegates saying a succinct "Hello" in their respective languages, but delegates stepped up to the microphone with much longer speeches in mind. Some read poetry, others offered inspirational quotes, and all appealed to a sense of global unity on behalf of their individual nations. And who could really blame them? The likelihood of alien interception was minuscule, but a worldwide terrestrial audience was guaranteed, a rare opportunity to invigorate the spirit of humanity. Time limits proved no match for politicians, so Sagan instead turned to the foreign language department at Cornell University to complete the short well-wishes.

There was one UN message the team found particularly poignant, though, and an audio greeting offered from Secretary General Kurt Waldheim was included on the record in full: "As the Secretary General of the United Nations, an organization of 147 member states who represent almost all of the human inhabitants of the planet Earth, I send greetings on behalf of the people of our planet. We step out of our solar system into the universe seeking only peace and friendship, to teach if we are called upon, to be taught if we are fortunate. We know full well that our planet and all its inhabitants are but a small part of the immense universe that surrounds us and it is with humility and hope that we take this step."

It was a beautiful sentiment, worthy of inclusion, but politics had officially entered the game, and NASA was a savvy player. In a move sure to confuse any alien who came across it, the records also contained the names of members of the US Senate in 1977, a token of gratitude for their mission funding support. And if they were making space for the senators and the UN Secretary General, it was only fair to offer the same opportunity for the president of the United States. Perhaps surprising from a world leader, President Jimmy Carter was very considerate of the space limitations. His inspirational greeting was concise, and he offered it in text rather than audio:

"This is a present from a small distant world, a token of our sounds, our science, our images, our music, our thoughts, and our feelings. We are attempting to survive our time so we may live into yours. We hope someday, having solved the problems we face, to join a community of galactic civilizations. This record represents our hope and our determination, and our good will in a vast and awesome universe."

By now, dramatic feats of space exploration had inspired imaginations around the world of all ages and genders. And although the terminology at the time was "manned spaceflight," women already had a sole representative. Back in 1963, Soviet Union cosmonaut Valentina Tereshkova had piloted *Vostok 6* and become the first woman in space, a source of gender equality chagrin for the United States, whose early space leadership took a more paternalistic point of view.

The year before Tereshkova's flight, a young girl sent a letter to President Kennedy offering her own future service in America's astronaut corps. NASA wrote back, "Your willingness to serve your country as a volunteer woman astronaut is commendable. However, while many women are employed in other capacities in the space program—some of them in extremely important scientific posts—we have no present plans to employ women on space flights because of the degree of scientific and flight training, and the physical characteristics, which are required." Ouch.

Cracks had already begun forming in the pillars of that argument though, and at the same time the little girl received her rejection letter back in the mail, one woman was storming through the halls of Congress to petition the rules. NASA hadn't set out to intentionally exclude women from the astronaut selection process, but as it turned out, the candidate criteria would preclude them. Early on, NASA had reasoned the ideal astronaut candidates would need to be well-acquainted with stress and adrenaline. They should be capable of handling emergencies under immense pressure, and ideally, they would have experience with the physical and psychological stresses of high-speed flight.

All potential candidates with those baseline qualifications would then need to undergo extensive medical testing to ensure their bodies could withstand the rigors of spaceflight. From the beginning, NASA knew the program would be incredibly popular, and there was no way they would be able to medically evaluate the avalanche of qualified-on-paper applications they were certain to receive. Instead, they brainstormed a shortcut that seemed to solve the issue: if they limited their search to military test pilots, they could be sure all candidates had the right temperament for experimental flight and work from a much more precise, pre-vetted candidate pool.

And so, the initial candidate qualifications were decided: all astronaut applicants must be graduates of military test pilot schools. And in that decision lay the disqualifying issue for half of the American population: at that time, military test pilot schools were male-only institutions.

For any test pilot interested in becoming a NASA astronaut, the next step in the selection process was a battery of invasive medical tests, physical endurance challenges, and mental evaluations. In between constant medical monitoring and psychiatric interviews, they raced on treadmills, tolerated electric shocks, endured prolonged periods of sensory deprivation, and blew up balloons to the point of exhaustion. Space was uncharted territory, and beyond perfect health, scientists weren't quite sure what performance thresholds the new environment might demand. Through difficult elimination, NASA selected their first seven astronauts, dubbed the Mercury 7.

As America's newly minted astronauts adjusted to their new careers and public interest, NASA flight surgeon and head of Life Sciences Randy Lovelace took a fresh look at the data. Together with aviation pioneer Jacqueline Cochran, he wondered whether the typically smaller and lighter frames of women might be even more ideal for spaceflight. Using his private clinic and financing from Cochran and her husband, Lovelace set out to put a group of female candidates through the same rigorous testing to see how they fared.

To mirror the official NASA astronaut selection process as closely as possible, Lovelace limited testing invitations to accomplished female pilots with more than one thousand hours of flight experience. From these First Lady Astronaut Trainees (the FLATs), Geraldyn "Jerrie" Cobb became the first woman to undertake and pass all the same tests as the Mercury 7, even outscoring nearly all of her male counterparts. Twelve more women would follow in her footsteps, and together the exceptional women would be dubbed the Mercury 13.

Armed with the proof of her own accomplishments, Cobb helped force the issue of gender equality all the way up to Congress, where she was called to testify in a hearing in front of the Special Subcommittee on the Selection of Astronauts. Cobb laid out a compelling and data-driven case for the inclusion of women, but not everyone was convinced. Mercury 7 astronauts John Glenn and Scott Carpenter were also called to testify, and while they agreed with the capability of women in theory, they stopped short of advocating for their inclusion in practice. "It is just a fact," explained Glenn, "The men go off and fight the wars and fly the airplanes and come back and help design and build and test them. The fact that women are not in this field is a fact of our social order. It may be undesirable. It obviously is, but we are only looking, as I said before, to people with certain qualifications. If anybody can meet them, I am all for them."

But until either NASA adjusted their criteria or military test pilot schools opened their doors to women, those certain qualifications would remain eligible only to men (even though, ironically, an exception had been made for Glenn himself, who did not possess the requisite college degree at the time of his astronaut candidate selection). The stalemate would last another seventeen years until NASA made the first move to broaden their pool of talent.

By 1978, NASA was ready to announce a new group of astronaut candidates, the first since the moon landings. In addition to mapping out blueprints for a new orbiting space shuttle, the space agency had also finally figured out how to diversify talent in the astronaut corps. In addition to pilots with the traditional backgrounds, Astronaut Group 8 would also recruit for the brand-new role of mission specialist. This opened the door to a host of new expertise, including the first six female astronauts: biochemist Shannon Lucid, physician Margaret Rhea Seddon, engineer Judith Resnik, geologist Kathryn Sullivan, physicist Sally Ride, and chemist Anna Fisher, who would later become the first mother in space. The historic cohort also included the first three Black astronauts, physicist Ronald McNair and pilots Frederick Gregory and Guion Bluford, along with engineer Ellison Onizuka, America's first Asian astronaut.

By 1983, women were finally gearing up for spaceflight. Sally Ride had her flight assignment on the new space shuttle *Challenger*, along with the distinction of becoming the first American woman in space. But while Sally was focused on her science, American media insisted on steering the conversation toward her gender. Journalists wanted to know about her family plans, whether she cried when things got stressful at work, and even about the undergarments she would wear in space. Late night comedians doubled down on the stereotypes, with Johnny Carson joking that the shuttle launch was going to be postponed until Sally could find a purse to match her shoes.

Even within her own team, Sally was constantly reminded of the novelty in a woman's participation in spaceflight. A new privacy curtain was installed around the shuttle's vacuum toilet and well-meaning engineers reportedly wondered aloud whether one hundred tampons would be the right number to pack for a seven-day flight. (It would not be the right number.) Although far from seamless, Sally's integration paved the path for

American women in space, and "manned spaceflight" gave way to "human spaceflight."

The space shuttle program that carried Sally was unique in that the orbiter itself was reusable. Strapped with two rocket boosters for takeoff, the vehicle was capable of gliding to a landing back on Earth, to be launched again with new rocket boosters. The result was a fleet of five orbiting laboratories that, over the course of thirty years and 135 missions, would carry 355 astronauts to space to conduct research, service space-based equipment, and launch experiments. But the incredible achievements of the space shuttle program were also punctuated with tragedy, and through the catastrophic losses of space shuttles *Challenger* and *Columbia*, America learned more hard lessons about the perils of pushing the envelope and the costs of opening new frontiers.

The tenth flight of the space shuttle *Challenger* was set to be a special one. Along with Commander Dick Scobee, pilot Mike Smith, payload specialist Gregory Jarvis, and mission specialists Judith Resnik, Ellison Onizuka, and Ron McNair, the orbiter was carrying the first ever civilian selected for spaceflight: a schoolteacher named Christa McAuliffe. McAuliffe had been selected from more than 11,000 applicants to participate in NASA's new "Teacher in Space" project, earning her the opportunity to inspire students on an ultimate field trip.

Classrooms across the nation were tuned in to their televisions on January 28, 1986, watching live as McAuliffe and her *Challenger* crewmates boarded the space shuttle on an unusually chilly morning in Florida and took off for space. But just seventy-three seconds into flight, space shuttle *Challenger* broke apart in the air, killing all seven crew members and disintegrating over the Atlantic Ocean as nearly 20 percent of the US population watched in horror.

The carnage would later be traced back to a design flaw in the shuttle's rubber O-ring seals, which weren't equipped to operate in the cold temperatures, and whose failure initiated a complete structural collapse. The nation grieved and NASA hit pause on the shuttle program, dedicating nearly three years to investigations, redesigns, and procedural overhauls aimed to prevent such a disaster from happening again.

Sadly, tragedy would strike again seventeen years later, as Commander Rick Husband, pilot William McCool, payload specialist Ilan Ramon, and mission specialists Michael Anderson, David Brown, Kalpana Chawla, and Laurel Clark were preparing to return from nearly sixteen days on board space shuttle *Columbia*. After two weeks of conducting research and carrying out dozens of experiments, it was time for the crew to come home, but back at Mission Control, some engineers were beginning to worry.

During the initial launch two weeks earlier, video feeds revealed a piece of foam insulation had broken off from *Columbia*'s external tank and struck its left wing. The crew was aware of the debris, but NASA had studied the footage and determined that the damage wouldn't cause any issues on the way back, emailing the crew, "We have seen this same phenomenon on several other flights and there is absolutely no concern for entry." Some folks on the ground, however, weren't so sure.

As reentry day drew near, a few engineers wondered aloud about the worst-case possible extent of damage. If the Orbiter's thermal protection system had been compromised, there was no longer a chance at in-orbit repairs. NASA determined that the system should be fine, but some folks at Mission Control reportedly still grappled with the morbid hypothetical of whether a doomed crew would want forewarning. NASA remained optimistic; foam had flown off before and it hadn't been an issue; they believed this time would be the same.

But on February 1, 2003, as space shuttle *Columbia* reentered Earth's atmosphere at twenty-five times the speed of sound, the heat shield failed. With its internal wing structure crumbling, the spacecraft depressurized and broke apart over the state of Texas, killing all seven crew members on board.

Once again, the space shuttle program was suspended while inquiries were made, procedures were overhauled, and contingency plans were enacted for future flights. Mistakes had been made and lives had been lost; America's future in space seemed uncertain but the underlying reality had never been clearer: spaceflight was incredibly hard and incredibly risky.

This was a turning point: as the husband of fallen astronaut Laurel Clark explained at the *Columbia* crew's memorial, America could be space-fearing or spacefaring, but to continue to move forward there would have to be a national acceptance of risk. "This memorial has many blank spots, and they will not go unfilled," he continued, "because the destiny of mankind will come at some cost." And in spite of the tragedy, President George W. Bush addressed a mourning nation with a similar promise of resilience, committing, "The cause in which they died will continue... Our journey into space will go on."

🚀🚀🚀

Although America's space shuttle program would only last another eight years after the *Columbia* tragedy, those later missions enabled lightyears of scientific advancement, and helped reveal how much we still have to learn about the universe around us.

Until the twentieth century, it was assumed that our Milky Way galaxy made up the entirety of the universe. And looking through a telescope, it appeared to be a dusty universe, filled with smudges that blurred the view for astronomers like Edwin Hubble. But with the help of Mount Wilson Observatory's new Hooker Telescope, the most powerful in the 1920s, Hubble was able to take a fresh look at the night sky. He discovered stars whose pulsating brightness provided a method for calculating their distance, a measurement he realized was much too far away to be a part of our own galaxy. That meant those hazy clouds were not merely clusters of gas and dust at the edge of our own galaxy—Hubble's observations revealed that each of those little smudges were galaxies of their own, strewn across a universe that was—wait for it—expanding in size.

In a complete astronomical mic drop, we had jumped from the assumption of one galaxy to the knowledge of more than a billion, and we were going to need much more powerful telescopes to study them. Seventy years and several prototypes later, the aptly named Hubble Space Telescope would be packed into the cargo bay of space shuttle *Discovery*, on its way to low Earth orbit to help expose the deepest corners of the universe.

Those iconic images would reveal a treasure trove of scientific data that galvanized astronomers and the public alike. Hubble would help us nail down a dizzying array of discoveries, like a more precise age estimate of the universe (13.8 billion years), supermassive black holes (not only do they exist, but most galaxies have them), and fascinating geological features of the moons of planets like Jupiter (Ganymede has an ocean! Europa has plumes!)

The scale of discovery was staggering: hundreds of billions of galaxies out there and we hadn't even fully explored the corners of our solar system, let alone the rest of the Milky Way. Earthlings had a lot left to discover, and where better to start than our closest planetary neighbor?

There has always been something uniquely compelling about the red planet, our next-door neighbor a world away. From *The War of the Worlds*, to *The Martian Chronicles* and *Stranger in a Strange Land*, Mars has long captivated imaginations, offering a siren song for exploration and teasing the possibility of other life in the universe.

The journey to Mars began as a space race in its own right and getting there wasn't easy. Mars is a notoriously difficult destination to reach, with a mission success rate hovering just under 50 percent. Even though the *Mars 2* and *Mars 3* probes failed upon reaching the Martian surface (like poor Albert, *Mars 1* never even made it that far), it was still a win for the Soviet Union, who had managed to send the first human-made objects to Mars. America was close behind, and as with previous space competition, the US played to win.

NASA was finding success with their Mariner probes and when *Mariner 9* became the first spacecraft to enter Mars's orbit, it returned photos that changed our perception of the red planet and encouraged our wildest space dreams. We learned that at some point in the past, liquid water likely flowed on the surface of the rocky planet, and we saw for the first time the full size of Olympus Mons, which at two and a half times the height of Mt. Everest was immediately reclassified as the tallest volcano in the solar system.

Now that NASA had indications of water, they were ready to search for indications of life, and in 1975 *Viking 1* and *Viking 2* launched in hot pursuit of those coveted biosignatures. The data was inconclusive, which in scientific terms is to say it was wildly encouraging. But familiar obstacles quickly emerged: we needed more time and more money.

Two decades and several significant technological advancements later, we had just the robot for the job—a cost-effective rover named Sojourner. After a successful delivery from robotic spacecraft *Pathfinder*, Sojourner became the first rover to spin its wheels on the surface of another planet. Designed for a seven-sol mission, the rover managed to weather eighty-three sols (or eighty-five Earth days) and the data sent home provided further evidence that our cold, dusty neighbor was once a much warmer, wetter planet. *Pathfinder* and Sojourner also served as proof-of-concept for the follow-on missions NASA envisioned, including a series of rovers who would eventually follow in Sojourner's wheel-tracks and capture hearts and minds the world over.

First up was Spirit, one of two golf cart-sized rovers tasked with a ninety-sol mission to explore areas where liquid water—and perhaps even life—may have flowed in the past. Dust storms on Mars are deadly for rovers, reducing their mobility and slowly burying the solar panels they depend on for power. Such was initially the fate of Spirit, until a group of dust devils swirled past the rover, blowing the regolith clean off of its solar panels and increasing power to more than 90 percent. Spirit soared back to life. Dirt would still be the rover's demise though. By the time Spirit came to its final resting stop in 2010, its wheels stuck in soft sediment, the rover had exceeded its expected lifespan by more than twenty-five times, clocking six years, two months, and nineteen Earth days.

Spirit's twin was an even bigger overachiever. Opportunity, nicknamed "Oppy," knew exactly how much destruction a dust storm could wreak on a rover and managed to navigate shelter and repeatedly harness solar power for a shocking fourteen years and forty-six Earth days past its designed lifespan. In the final transmissions before Oppy went quiet, the rover sent back data warning of diminished power and volatile weather. It was unlikely that sunlight could penetrate the storm to help the rover recharge, and as NASA launched a commendable effort to re-establish contact, the #OppyPhoneHome hashtag trended across the nation on Twitter. One writer paraphrased the Opportunity's last transmissions so poetically that it had some Earthlings in tears (okay, fine, just me) over the beloved rover's "final words" as it powered down on another world: "My battery is low and it's getting dark..."

And because the depths of our ability to anthropomorphize know no bounds, Opportunity's successor Curiosity became an instant and verifiable celebrity before it even left Earth. The rover was the most advanced chemistry lab ever sent to Mars, but it was Curiosity's personality that really struck a chord with the public. In addition to groundbreaking science, the rover was prone to selfies, tweeting in the first person, and occasional moments of sentimentalism. Back in 2013, thanks to the clever programmers at JPL, Curiosity hummed "Happy Birthday" to celebrate the end of its first year on Mars.

What a time to be alive, when science fiction fans could finally point to Mars and say with scientific certainty: that right there is a planet inhabited entirely by robots. But as the name of the initial spacecraft *Pathfinder* indicated, these robots were always intended to precede human exploration, never to replace it altogether. The secrets of the cosmos were alluring, and the worthiness of expanding human presence in space was a cause that multiple countries could get behind.

After the Space Race, competition dissipated into business cooperation, with the United States and Russia both bringing other nations along for the ride. Russian Soyuz spacecraft and American space shuttles served as cosmic ride-shares, ferrying government-sponsored astronaut representatives from dozens of different nations, many of whom had no national space programs of their own. The arrangement was transactional, but the cooperation was exemplary, and it set the stage for the world's most impressive joint venture. The International Space Station would be a football field-sized orbiting laboratory and a home away from Earth for astronauts and guests of five different space agencies: NASA (United States), Roscosmos (Russia), JAXA (Japan), ESA (Europe), and CSA (Canada).

While not the first space station, the ISS can certainly claim the title for most productive. In fact, thanks to the ISS, anyone born after November 2, 2000, has never known a time when humans weren't living and working in space. That's the date the first long-term crew moved in, and the Station has enjoyed rotating but permanent occupancy ever since.

From 250 miles above the Earth's surface, astronauts on the station have leveraged the microgravity environment to conduct an impressive range of scientific research across a number of fields, unlocking a long and varied list of earthly benefits and capabilities. Among many other spin-off applications, high-definition images captured from the space station contributed to disaster relief efforts back on Earth; biological data logged from long-term crew advanced our understanding of medical concerns like bone density loss and osteoporosis, and by virtue of the space station's remote location, the fields of robotic surgery and tele-medicine were propelled into the future.

More than two hundred astronauts have floated through the lab, and while scientific research has always been at the forefront of its mission, the ISS has also served as a diplomatic bridge when earthly relations were taut, and yes—even a coveted tourist destination.

The ISS became our temple of space, a feat of engineering that proved our capability and worthiness as a spacefaring species. Continuous occupancy evidenced our commitment to human spaceflight, proof that our previous jaunts out of Earth's gravity were more than just stunts or flukes. On any given night, people on Earth could look up and track the orbiting lab across the sky, marveling at how far we've flexed our talents. Above all, it was a testament to the millennia of hard work that got us to this point as a species.

There are certain moments that have the power to change the course of history. These moments are giant leaps, unlocking new potential for a species that can chart their own course of evolution. One of those giant leaps occurred more than three million years ago when the group of *Australopithecus afarensis* stood upright and left footprints across the volcanic ash. Another giant leap occurred closer to 200,000 years ago, when *Homo sapiens* first distinguished themselves as a species, and in the centuries that followed, spread their footprints across new continents. And another giant leap occurred in 1969 AD, when *Homo sapiens* launched from Earth and left footprints in the moon dust. It makes you wonder what this spacefaring species is capable of accomplishing next. If we're still around a few millennia from now, it means we've mastered space travel and life off Earth, and we've likely achieved a couple more giant leaps along the way.

Perhaps next up is an Age of Adaptation, in which *Homo sapiens* unlocks the bio-logistics of settling a new planet or space colony. Future space-adapted historians in those settlements will look back on our early Space Age and see a movement painted in broad strokes. Like each age before, the details will be lost to the general momentum, a brief chapter summary in a linear tome of human progress. Of course, living through this Space Age ourselves, we know the path feels far from linear, and everything depends on the details. It's in these details that we find the agency to chart our path and influence the future. But to have our details one day summarized into cosmic CliffsNotes is to have successfully passed the baton—it means we not only survived, but we teed up another great age for our descendants. It's difficult to fathom a more noble purpose.

As the Apollo program reminded the world, in addition to boldness and innovation, giant leaps require fierce determination and a whole lot of money. We had figured out the enormous engineering challenges of human spaceflight only to struggle with the economic ones, and there was a growing frustration that, decades after landing on the Moon, we hadn't been back, let alone traveled further. Human spaceflight felt stagnant. We needed a revival. We needed an encore. As we know, the survival of our species has always depended on a multitude of talents and contributions, so who better to take a fresh look at the government's billion-dollar problems than the billionaires themselves? Space caught the attention of some of the most talented entrepreneurs on the planet, and damned if they were going to let the baton drop on their watch.

CHAPTER 3

RISE OF THE COMMERCIAL SPACE INDUSTRY

MUCH TO MY CHAGRIN, I WAS A LATE-BLOOMING SPACE NERD. I WAS ALWAYS

fascinated by the idea of humans living and working in space, but it was a long time before it occurred to me that anyone might be able to take part, and far longer before I came to embrace my own place in the Space Age. Even now, when I don a spacesuit myself, I have a quick pinch-me pep talk with the part of myself who had always assumed that training for spaceflight was reserved only for those few who had proven their worthiness and aptitude across an entire lifetime of commitment—like prodigious kindergarteners who had really meant it when they said they wanted to be astronauts when they grew up, and charted a course from there.

Instead of deep career contemplation, the 2000s had been a decade of self-discovery for me. Figuring out who I was during middle and high school was hard enough without the pressure of deciding what I was going to do when I got older. While my more ambitious peers packed their trunks for Space Camp, I was being interviewed by my local newspaper as part of a profile on kids who spend too much time playing video games. Worse, I was self-nominated. Given my talent for punching buttons on a controller, the reporter wanted to know what I had in mind in terms of a career. If there was sarcasm in the question, it went right over my tween head. I answered earnestly and truthfully that if things continued to trend this positively for me, I hoped to one day land a job selling video games at the Electronics Boutique in my local mall. "She's on the honor roll," my mother is quoted as saying, "so we're not overly worried."

It would take me a few more years to sort out my specific career goals (and many more years of practice to articulate them to the press), but space had always been lingering in the background of my life. I read the books, watched the movies, and of course, played the video games. And if the entertainment exposure to the excitement of space wasn't enough, I even witnessed the real thing from the comfort of my own home in Jupiter, Florida. My bedroom window faced northeast, perfectly framing the stretch of sky over Cape Canaveral, and from the side of my bed I watched dozens of astronauts make the journey to space.

That my adolescent years were punctuated with space shuttle launches certainly helped nurture an appreciation for the wonder of spaceflight, but there was still a governor limiting my imagination. I had compartmentalized space travel as something a handful of incredible humans were born to do and watching them fulfill their dreams was simply a cool perk of being alive in the twenty-first century.

I hadn't paid much attention to how we got there, and until the Columbia disaster, I hadn't given too much time to why we went in the first place. Two weeks earlier, my class had filed on to the football field to watch *Columbia* take flight. Back inside, we had chatted lightly about the research they would do and covered the usual highlights of everyday chores and personal hygiene in space. I knew the kinds of food they would eat, how they bathed, and even how they went to the bathroom up there, but nothing about them as people, or how they had navigated a career that would take them to space in the first place.

On a Saturday morning two weeks later, I woke up to the shock of tragedy. The news was full of technical details I didn't understand, but with the faces of *Columbia*'s crew and their family members plastered all over the screen, I saw them for the first time as people. These weren't just astronauts. These were husbands, wives, fathers, daughters, and—this last shock stuck with me in ways it would take years to unpack—mothers.

At some point my mom turned off the news, but not before I watched President George W. Bush address a grieving nation. His words spoke directly to the root of distress and confusion I was feeling. "In an age where space flight has come to seem almost routine," he explained, "it is easy to overlook the dangers of travels by rocket, and the difficulties of navigating the fierce outer atmosphere of the Earth. These astronauts knew the dangers, and they faced them willingly, knowing they had a high and noble purpose in life."

The speech ended with an affirmation that our journey in space—the cause in which they died—would continue. From that moment on, I looked at space travel through a slightly sharper lens. That we would willingly confront this peril for a higher purpose fascinated me. I wanted to understand the motivations of everyone involved in this quest for discovery.

I don't recall the exact moment it dawned on me that there was an entire workforce enabling the flights of the space shuttles I watched soar through the skies, but the understanding that astronauts were just the tip of the iceberg adjusted the limiter on my imagination. There were tens of thousands of people out there dedicating their lives to laying the groundwork for human space exploration; I was finally beginning to find myself in the conversation, not as a space expert, but as a fellow *Homo sapiens* with skin in the game. It would take a few more years for my passion to go full throttle, but I was motivated to investigate why the advancement of human spaceflight was such a worthy purpose.

Meanwhile, that space-enabling workforce was expanding. While I was still catching up on all of our accomplishments in space, more creative minds were already imagining how much more we could achieve. Unlocking the final frontier wasn't just a thrill, it was a long-term necessity. Entrepreneurs were looking at space travel and calculating all of the ways to make it better, faster, cheaper, and best of all, more accessible. While I was playing Xbox, my future colleagues were laying the foundation for the commercial spaceflight industry.

Of course, large aerospace companies like Boeing and Lockheed Martin (and later, their United Launch Alliance) had long served America's space and defense program, developing exactly what NASA and other government agencies asked of them and racking up huge government contracts in the process. These heritage institutions had delivered on some of the world's most significant achievements, but in the process had earned the reputation of being bureaucratic and slow-moving, and a lack of competition meant no real business incentive to lower the cost of launch or risk success with new approaches. But in the cradle of Silicon

Valley, new companies were emerging to disrupt the old way of doing business, turning their attention to increased agility, speed, and innovation that would dramatically lower the costs and open access to space for everyone.

While many companies focused on creating new infrastructure for space activities, some were figuring out how best to leverage what was already there. By the end of the International Space Station's assembly, a company called Space Adventures had already realized the lab's potential for hosting private space travelers, an idea that wasn't entirely without precedent.

NASA had already flown a select handful of international astronauts and American corporate contractors on the space shuttle. In fact, the role of "payload specialist" had been created specifically to accommodate those who had been selected and trained outside of NASA's normal astronaut selection process. The payload specialist role had opened the door for a slightly broader but still limited pool of people, including representatives from partner nations, politicians, and civilian contractors tending to specific payloads (such as an engineer assigned to fly alongside their corporation's communications satellite).

For Space Adventures, the International Space Station represented an opportunity to further expand that access to private citizens and prove that low Earth orbit could be a viable industry. To make the business case, they needed to identify a unique population of civilians who could both afford to self-fund a mission to space and who had enough interest to commit to training for one.

Dennis Tito sat right in the intersection of that Venn diagram. A former engineer and scientist at NASA's Jet Propulsion Laboratory, Tito had applied his mathematical skills to the field of finance with great success. He redirected his analytical talents from spacecraft trajectories to market risks, and in doing so helped to develop the financial field of quantitative analytics.

Throughout his incredibly lucrative career in investment management, Tito still held a lifelong passion for space exploration and a dream of traveling there himself. Roscosmos had already demonstrated eagerness to engage in commercial activity, and in partnership with Space Adventures, a twenty-million-dollar deal was brokered with Moscow to fly Tito on an upcoming Russian Soyuz flight. At sixty years old he felt like the time to try was now or never. However, some officials at NASA felt like things were moving too fast.

Tito trained alongside two Russian crewmates for nearly a year at the Cosmonaut Training Center in Star City, Russia. But when the trio showed up to NASA's Johnson Spaceflight Center to continue with US-based training, NASA officials wouldn't allow Tito to participate, citing a tangle of liability and safety concerns that still needed to be unraveled. Disagreement between NASA and Roscosmos was tense. When the Americans expressed concern that his previous training wasn't sufficient for operating in the American modules of the space station, the soft-spoken Tito reportedly remarked that he would be quite content to stay on the Russian side.

In April 2001, in spite of NASA's misgivings, Tito launched aboard *Soyuz TM-32* and spent nearly eight days on the International Space Station, orbiting the Earth and conducting science experiments. His flight represented a watershed moment for commercial human spaceflight. In partnership with Roscosmos, Space Adventures would shepherd six more private astronauts into orbit over the next decade. Together with Dennis Tito, entrepreneurs Mark Shuttleworth, Greg Olsen, Anousheh Ansari, Charles Simonyi, Richard Garriott de Cayeux, and Guy Laliberté shared a passion for space exploration and a desire to use their great business success to help open the space frontier for civilians. Their self-funded flights marked the beginning of a new space tourism industry, but to call them tourists would minimize their preparation and achievement. Their orbital training was extensive, covering more than eight hundred hours of instruction on emergency procedures, life-support and engineering systems, simulations of spaceflight and microgravity, as well as preparation for individual scientific and research goals. On an adventure-tourism scale, this was a Sherpa-guided summit of Mt. Everest, not a bus tour of Machu Picchu.

Just as Jerrie Cobb had been called to testify to Congress about the capabilities of women in space, so too was Dennis Tito asked to submit testimony on behalf of private citizens. In a Joint Hearing on commercial human spaceflight, he reinforced how well his training had prepared him and praised the capabilities of the ISS as a platform for scientific research and inspiration. He had shown that civilians were capable of sharing in America's space program, but he also wanted to drive home that they were worthy of it.

"It is hard for me to fully convey what it was like to be weightless for eight days. But then again, I'm a businessman. On the other hand, just think of how magnificently poets, writers, musicians, composers, teachers, filmmakers, painters, journalists, and other creative individuals would be able to communicate the beauty and inspiration of spaceflight."

Of course, not everyone had twenty million dollars to spare for spaceflight. If we were going to find a way to open up space to everyday people, we needed to lower the cost of launch. And in a quest for affordability, reusability was the holy grail. The expensive rockets that propelled a spacecraft were designed for single use, the equivalent of throwing away an airplane and rebuilding it after each international flight. Recycling a rocket and servicing it between flights would mean huge savings on cost to orbit. The partially reusable space shuttle had been a step in this direction, but the delicate systems and extensive refurbishing required between flights had made it even more expensive than expendable rockets. This was a perfect problem for the commercial spaceflight industry to solve—the industry just needed a jump-start.

The concept of reusability was foremost in mind for entrepreneur Peter Diamandis. In the late 1990s, he reflected that prize money had long served as starter fluid for innovation; seventy-five years earlier, hotelier Raymond Orteig had put up $25,000 to inspire aviators to take flight across the Atlantic Ocean, a prize claimed by Charles Lindbergh and his famed *Spirit of St. Louis*. With that success in mind, Diamandis dreamed up the X Prize, a ten-million-dollar prize for the first private organization to launch humans into space twice within two weeks. The repeat flights would demonstrate the innovation required to bring down the cost of launch, and a rule prohibiting government funding would further prove the capability of a commercial industry to build a business case around the challenge.

Opening up access to space for civilians and scientists was a goal that resonated deeply with Anousheh Ansari and her brother-in-law Amir. Ahead of Anousheh's own flight to space with Space Adventures, the two agreed to sponsor the contest, which was officially renamed the Ansari X Prize. Fully funded, the race was on to demonstrate that private industry was capable of building vehicles to serve future commercial and tourist markets.

The opportunity to innovate the spaceflight industry had attracted the attention of other entrepreneurs as well. Aviation maverick and Scaled Composites founder Burt Rutan had a design for a reusable spacecraft named *SpaceShipOne*, which would be launched in-air with the help of a jet-powered mother ship called *White Knight*. Bold plans required bold investment, and Rutan found financial backing from none other than Microsoft cofounder Paul Allen.

Sure enough, *SpaceShipOne* had the right stuff for suborbital human spaceflight, and the regulatory bodies had to race to keep up. In June 2004, California's Mojave Airport received a space launch license from the Federal Aviation Administration, and just days later, test pilot Mike Melvill took to the skies in *SpaceShipOne*, becoming the first commercial astronaut of a private spacecraft and drawing thousands of spectators to the newly named Mojave Air & Space Port.

The successful test flight was encouraging, and just a couple of months later the team was ready to make a bid for the Ansari X Prize. In late September that same year, Mike Melvill flew again and completed the first of the back-to-back flights. Days later, on October 4, pilot Brian Binnie flew *SpaceShipOne* on its second flight to space within two weeks, making history and clinching the ten-million-dollar prize.

By all measures, the Ansari X Prize was a big success. In addition to proving the abilities of private industry, the contest had attracted more than two dozen teams around the world and tens of millions of dollars had been invested in pioneering reusable suborbital technology. Even better, the contest had identified a market in the excited public, many of whom were rekindling their own personal dreams of spaceflight and realizing for the first time that those goals might actually be achievable in their lifetimes.

SpaceShipOne's history-making flight also caught the attention of Sir Richard Branson, long known for his bold brand portfolio and creative approach to business. Inspired by the potential, Branson formed a new company within the Virgin Group. Virgin Galactic would offer space tourism flights, positioning themselves as the vanguard of a new industry that would build on the technology of *SpaceShipOne* to open up space to everyone. Virgin Galactic would partner with Burt Rutan and Paul Allen's team to form the Spaceship Company, the organization that would manufacture that next generation commercial spacecraft.

Branson confirmed what Diamandis suspected: a space tourism market existed. With a reusable spaceplane, Virgin Galactic intended to lower the cost of access to space by an order of magnitude, and for an initial ticket price of just $200,000, passengers could guarantee themselves a seat on a future ninety-minute flight to space. The response was overwhelming.

The day *SpaceShipOne* won the Ansari X Prize, Brian Binnie became the 435th person ever to travel to space. Just a few years later, more than 500 people had put money down for seats on Branson's suborbital spaceplane. The ticket sales foreshadowed the power of a robust commercial spaceflight industry—once a company like Virgin Galactic started commercial operations, they had the potential to singlehandedly double the number of human beings who had ever been to space.

Paul Allen and Richard Branson weren't the only billionaires looking up toward the stars. While Branson focused on enabling humans to tour the cosmos, PayPal cofounder Elon Musk was focused on getting them there to stay. The red planet had long captured his attention, and it shocked him to discover that, by the turn of the century, NASA had not yet put out a definitive timeline to send humans to Mars. Musk took matters into his own hands, brainstorming missions that might reenergize the world's interest in space travel. If he could get his hands on cheap, surplus Russian rockets, he reasoned he could refurbish them to send life to Mars, perhaps a colony of mice or a greenhouse full of plants.

After a frustrating trip to Moscow, Musk decided he could build the rockets himself, and much more efficiently. To advance our footprint in the solar system, we were going to need giant, reusable, reliable rockets. And just a few months later, Space Exploration Technologies Corp., or SpaceX, was founded to make it happen. With a long-term plan for reusable rockets that would carry the species to Mars, Musk assembled a team of brilliant engineers who shared both his vision and work ethic. The goal was audacious from the beginning; before they made it to Mars, SpaceX's liquid-fueled rockets would need to succeed in low Earth orbit, something that had never before been done by private industry. The challenge would be expensive and the odds uncertain, but to succeed would be to transform the spaceflight industry. SpaceX got to work on *Falcon 1*, a rocket named after the Star Wars *Millennium Falcon*, followed shortly by *Dragon*, a spacecraft whose name paid homage to whimsical song "Puff the Magic Dragon"—a cheerful nod to all the skeptics who considered his vision impossible.

While Virgin Galactic and SpaceX were making a splash in the media, another billionaire-backed company was quietly designing launch vehicles and rocket-propulsion systems of its own. Since his school years, Amazon founder Jeff Bezos had harbored big space dreams, some of which he had shared with his local paper after being named class valedictorian. In a 1982 interview with the *Miami Herald*, teenage Bezos laid out his plans to preserve planet Earth, describing colonies that could house the millions of humans who would one day live in orbit. Eighteen years later, after leading Amazon through a successful IPO and earning a coveted spot on the Forbes World's Billionaires list, Bezos was ready to put his money where his mouth—and heart—was.

This time around, Bezos offered far less detail about his plans. Like SpaceX and Virgin Galactic, Blue Origin envisioned opening the frontier of space to all through reusable rocketry, but the company was notoriously more secretive about their progress. In fact, the company flew so far under the radar that fans and journalists had to dig through public-record filings and run down rumors to gauge progress on *New Shepard*, the developmental space vehicle named in honor of America's first astronaut. If there was a new space race happening in the commercial sector, Blue Origin wasn't positioning itself visibly at the forefront. One of the few things shared with the public was the company's logo, and it was rich with symbolism. An elaborate coat of arms featuring two tortoises drove home their incremental approach to spaceflight, along with the motto "*Gradatim Ferociter*," Latin for "step by step, ferociously."

Billionaire backing was a huge help, but it wasn't a hard requirement to set up shop in the space industry. In the rain shadow desert of Mojave, California, a vibrant rocket garden was blooming, and at the turn of the twenty-first century, the

Mojave Air & Space Port was home base for a full-blown space renaissance. Amidst all the envelope-pushing activity, a giant test vehicle from defunct company Rotary Rocket sat on display in front of the spaceport. To some it was a source of inspiration; to others, the permanently grounded test vehicle was a warning about the inherent unpredictability of the space business.

Besides Scaled Composites, the Spaceship Company, and Virgin Galactic, Mojave's spaceport was home to a number of privately funded space efforts in the early 2000s. Some, like Armadillo Aerospace, had competed for the Ansari X Prize themselves and were further investing in their technology. Wide stretches of empty desert provided the perfect landscape for hazardous rocket tests and engine experimentation. XCOR Aerospace was also focused on building a suborbital spacecraft to support space tourism, and their EZ-Rocket was the first privately built rocket-powered airplane, a bespoke platform to test the propulsion technology that would power their later spacecraft designs. Other companies had ambitions to fly technology rather than people. To that end, Masten Space Systems was developing a line of vertical takeoff, vertical landing (VTVL) rockets to serve as technology test beds, robotic landers, and research platforms.

Beyond Mojave's hub of activity and sonic booms, private space companies were spreading out across the nation. Bigelow Aerospace was designing inflatable space habitats from Las Vegas and UP Aerospace was launching sounding rockets from New Mexico, where brand-new Spaceport America advertised themselves as the world's first purpose-built spaceport. Not all of these companies would last, but the US government was taking note of all the activity.

As the commercial space industry was ramping up, NASA's space shuttle program was ramping down. The planned retirement of the thirty-year program meant that America would no longer have the option to launch crew and cargo to the International Space Station from US soil. Instead, they would need to purchase rides on Russian spacecraft. A US government replacement could be many years away, so NASA needed a near-term solution to fill the gap in capability. The rise of the commercial spaceflight industry inspired the space agency to think creatively; after years of being the only player in low Earth orbit, NASA was realizing that they could help kick-start a market of orbital delivery companies and then hire those same companies to haul cargo to the International Space Station. A commercial market could free up NASA funds for their deeper space exploration objectives, and more pressingly, limit US reliance on foreign spacecraft to reach orbit.

NASA didn't offer prize money, but they certainly fostered competition. The Commercial Orbital Transportation Services (COTS) program allowed NASA to share institutional knowledge and invest resources in the private companies who could build orbital vehicles capable of making deliveries to the International Space Station. The catch? Competing companies would need to demonstrate upfront that they had the private funding required to make it happen. More than twenty companies submitted proposals and received rejections, but SpaceX and Orbital Sciences Corporation ultimately landed the coveted multi-hundred-million-dollar COTS contracts. And with the retirement of the space shuttle drawing near, they had their work cut out for them.

By 2008, SpaceX was hitting their stride and making good on their promises. After months of nail-biting, budget-draining launch failures, the company pulled off a feat only nations had managed before: *Falcon 1* became the first privately developed liquid-fuel rocket to orbit the Earth. A follow-up flight would launch a satellite into orbit, proving the achievement was not a fluke. SpaceX was here to stay, and the hard work paid off big time. With NASA's bet on private vehicle development proven out, the space agency was now ready to contract those private vehicles, splitting $3.5 billion between SpaceX and Orbital Sciences for future missions to ferry cargo to the space station.

Through partnership with private industry, NASA had helped usher in a new era of spaceflight. As always, though, the devil is in the details.

The early 2000s saw a space industry renaissance, an era of bold innovation and creativity in imagining new ways to make commercial human spaceflight a reality. Enlightenment was just around the corner though, and companies realized they would need to invest time figuring out how to make it all work within the blurry bounds of government agencies and regulatory bodies.

The commercial space sector had attracted significant investment, but regulatory question marks could quickly drive a company's profits into the ground, or worse, prevent them from ever lifting off in the first place. The Federal Aviation Authority's Office of Commercial Space Transportation had long been the authority on licensing rocket launches, but reusable suborbital launch vehicles like *SpaceShipOne* were a new breed. They didn't fit neatly into the FAA's rocket launch division, nor did it make sense to certify the experimental vehicles through the larger and more established commercial aircraft division. Thanks to a concerted effort from industry, Congress acknowledged the need for an update and in 2004 established a new FAA experimental permit process for developmental test vehicles like *SpaceShipOne*.

A reasonable launch licensing framework was materializing, but there were still a number of important details to iron out. If America wanted to promote the emerging technologies for commercial human spaceflight, regulatory frameworks needed to evolve alongside them, and the commercial space industry was going to need to work together to make their voices heard. An industry association provided the perfect forum to speak collectively and the Personal Spaceflight Federation (PSF) convened for the first time in 2005, when the leading group of commercial spacecraft developers, operators, and spaceports came together to provide the FAA with collective input on a number of proposed regulations addressing commercial human spaceflight.

Member companies like SpaceX, Scaled Composites, Space Adventures, Armadillo Aerospace, XPrize Foundation, XCOR Aerospace, Virgin Galactic, and the Mojave Spaceport were able to leverage their collective experience and agree on sensible regulatory foundations regarding safety, liability, and insurance issues for vehicles operating under experimental permits. Safety would always be the top priority, but some margin of flexibility would be necessary to develop and test experimental vehicles. That such a diverse membership could agree on any single issue was impressive, but the shared mission of regulatory reform motivated PSF come to consensus on dozens of recommendations, cementing a strong and unified industry voice for the commercial space industry.

By 2009, the commercial space landscape was growing exponentially, and many more companies and organizations had joined. Recognizing the diversity of businesses that existed in support of human spaceflight, the PSF rebranded as the Commercial Spaceflight Federation (CSF), cementing the group as the leading trade organization for a growing ecosystem of entrepreneurs creating high-tech jobs and driving billions of dollars of investment into the new markets of space tourism, orbital delivery, national security applications, and more.

By the end of the decade, CSF had further expanded, and members were punching far above their weight class. Bigelow Aerospace had launched two expandable space habitats into orbit, Virgin Galactic had debuted and tested *SpaceShipTwo* (christened "VSS Enterprise") and its mother ship *White Knight Two* ("Eve"), Masten Space Systems had won more than one million dollars demonstrating stable, controlled flight with their reusable VTVL rockets, XPrize Foundation and Google had launched a thirty million dollar Google Lunar X Prize to incentivize privately funded teams to land a robot on the lunar surface, and SpaceX's *Dragon* spacecraft was on track to become the first commercial craft to successfully rendezvous and attach to the International Space Station, demonstrating their ability to deliver cargo—and eventually crew—to low Earth orbit.

As the capabilities of the private space sector matched—and in some cases surpassed—those of legacy defense contractors, media headlines would describe the friction as a clash between "New Space" and "Old Space." The intense attention was perhaps the biggest indication that the commercial space industry was here to stay. And not a moment too soon: by the end of 2010, the space shuttle program was just months away from retirement, and I was months away from graduation, trying to figure out how a non-engineer could contribute to the world's most exciting industry.

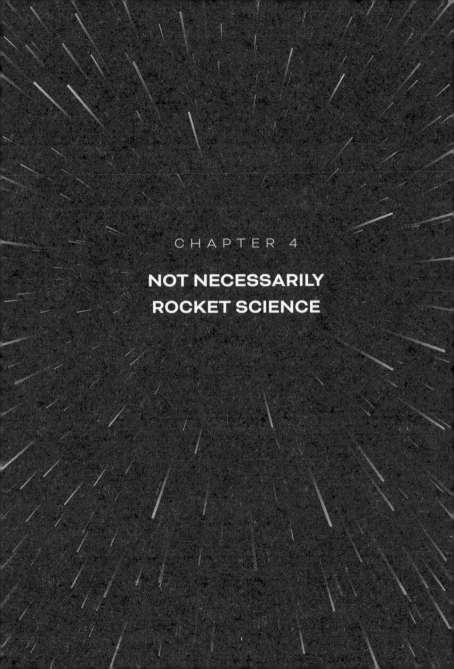

CHAPTER 4

NOT NECESSARILY
ROCKET SCIENCE

MY PATH TO THE SPACE INDUSTRY WAS INDIRECT, BUT I CREDIT THAT MEANDERING

route and resulting outlier status for my ability to create a unique impact once I got there. Folks are often shocked to hear that I earned a degree not in aerospace engineering but in film. The big reveal is particularly surprising (disappointing?) to the students who reach out looking for advice on selecting their college majors, attempting to reverse-engineer a career path in the space industry right out of high school. STEM degrees are indeed the norm, but they're by no means the requirement. If my years navigating the engineering world with a liberal arts degree have shown me anything, it's that professionals from the different disciplines of art and science have plenty to learn from one another. Still, I'm often presented with a hypothetical: if I could go back and do it over again, knowing what I know now, would I choose an engineering degree instead? Although I might have entered the space industry with a bit more context if I had, the answer is unequivocally no.

Perhaps naively, I approached college with an assumption that coursework in undergrad could and should exist entirely outside the scope of one's eventual career. I was excited to explore subjects I was completely unfamiliar with, and it was with that intentionally open mindset that I assembled an eclectic tasting menu of classes offered from Barnard College and Columbia University. I knew I liked space and science, but what else was out there? I adopted and eliminated new interests through trial and error. I balanced a fascinating anthropology seminar with an elective on the spaghetti Western genre of film (not a fan) and followed it up with a course on screenwriting (big fan). To fulfill my language requirement, I dedicated myself to Zulu, a Bantu language that incorporates clicks. Studying Zulu showed me that I could commit myself to learning and enjoying something on the absolute furthest edge of my radar and comfort zone.

I'll be the first to admit my choices were nonlinear, and to many, nonobvious. My father often wondered aloud when the last time was that I picked up a phone and had the opportunity to "press 2 for Zulu," but college was a rare opportunity to vet my interests and run them all the way down before contemplating a career path or graduate degree. Film, and more specifically the power of storytelling, had emerged as one of those interests, and I was eager to see where it could take me. I soon exhausted most of the electives available through Columbia's English department, where the study of film was only offered as a minor. Rather than declaring a different major, I transferred to New York University's Tisch School of the Arts, where a much more extensive catalog of courses was available to me. I tried my hand at cinematography and directing but it was the timeless art of storytelling that really drew me in; it clicked for me that an audience's attention was currency and the ability to capture it was an invaluable skill. The power of communication to inspire and motivate made me look at the world through a different lens.

A series of internships introduced me to the film and television industry, and while I greatly admired the production and development teams, I found my personal spike on the media and communications side. In addition to fetching coffee and temping as a front desk receptionist at a major film studio, I

NOT NECESSARILY ROCKET SCIENCE

had developed a reputation for quickly and skillfully drafting press releases for film acquisitions and award nominations. Soon I was tapped to help organize the media events for those films, and before long I was running them solo. I had no eye for filmmaking, but I could engineer a buzz-worthy limited screening. I was surprised and a bit disappointed to discover my film industry talents were more operational than artistic, but I found pride in my ability to deliver outcomes and make magic happen behind the scenes. I was soaking up knowledge and enjoying my work so much that I accepted an offer to return for a second year, despite the fact that I had already made up my mind that my broad interest in storytelling and human connection wasn't a perfect match for pursuing a career in the entertainment industry.

As far as I was concerned, the conclusion to my brief film career was no great loss. I was deeply appreciative to have had the opportunity to learn from titans of industry, even if that specific industry was one in which I no longer intended to work. In fact, I found my second year to be the most valuable, when I showed up every day not to further the goal of a future career there but to soak up every last bit of knowledge for a chapter of my life I was preparing to close out. I had staked my academic career on an industry which held no professional future for me, but I still felt like I had come out ahead. I learned that I'm the type of person who will go all in to see something through. Knowing with absolute certainty what I didn't want to do empowered me to focus my full attention on what I did want to do. Now all I had to do was figure out what that was.

Shortly after my graduation, space shuttle *Atlantis* touched back down on Earth for good and marked the retirement of the thirty-year space shuttle program. The news shocked me. This had apparently been in the works for years, but I hadn't been following it closely. I had grown up against the backdrop of space shuttle launches and I was deeply disappointed to see that era end; it seemed like a devastating setback for the nation. Worse, it seemed like a devastating setback for the nation. I wasn't yet familiar with the commercial spaceflight industry, so with that final landing of *Atlantis*, it seemed to me that America had willingly divested itself of NASA, foregoing the ability to launch astronauts to space from US soil. Worse, it felt like humanity's enthusiasm for space exploration was waning.

And yet, in other parts of the world I discovered the NASA brand was still going strong. Still digesting my film degree, I had set off on an embarrassingly predictable postgrad quest to gain a broader perspective. My own "Eat Play Learn" anthropological adventure took me to some of the most remote corners of the Earth. I tracked melting glaciers on a summit of Mt. Kilimanjaro, trekked up a waterfall to visit a remote Hmong village in Laos, and had honorary brass coils placed around my neck courtesy of the long neck Kayan women in Myanmar. I wanted to explore new environments and connect with humans whose life experiences differed vastly from own. I was expecting to see the world from a new perspective, but I was certainly not expecting to see my own country's space agency. And yet from

Tanzania to Thailand and everywhere in between I spotted NASA T-shirts, eventually making a game of finding the familiar blue logo in a crowd. The excitement and curiosity represented by that logo transcended all borders; NASA's bold pursuit of space exploration held the power not only to send people further out in the solar system, but also to bring them together right here on Earth. The realization that the pursuits and benefits of my country's space agency were global shook my perspective. I had set out to explore the Earth, and in doing so, I had discovered space.

Back home in New York City, I soon found kindred spirits in The Explorers Club, a century-old scientific society whose early members had carried the club's flag on an incredible series of "Famous Firsts," including the first expeditions to the North Pole, the South Pole, the summit of Mt. Everest, the deepest point in the ocean, and even the surface of the Moon. The Explorers Club Annual Dinner (ECAD) celebrated those who had dedicated their lives to discovery and pushed the boundaries of exploration, and it was there, at the Oscars of exploration, that I met my original mentors in the space industry, Laetitia and Richard Garriott de Cayeux.

In what remains the first and only practical application of my middle school video game obsession, I was immediately able to spot Richard from the across the room at ECAD. Except I knew him not as Richard but as "Lord British," the fictional ruler of Britannia from the video game franchise *Ultima*. I scuttled past archaeologists, mountain climbers, and oceanographers in a beeline for Lord British. I was so surprised to see the video game developer in person that I failed to notice he was standing in front of a spacesuit, no less one that bore his name.

It's worth noting that I was far from the only *Ultima* fan. Through the evolution of the series, Richard coined the term "massively multiplayer online role-playing game" (MMORPG), giving a name to the hobby into which I poured years of my adolescence. His successful career as a video game developer and entrepreneur enabled Richard to pursue his dreams of spaceflight, floating in the footsteps of his father, NASA astronaut Owen Garriott. Richard was a cofounder of Space Adventures and in 2008 flew on a self-funded mission to the International Space Station, returning to Earth as America's first second-generation astronaut. Laetitia was no stranger to space either. Just a few years away from cofounding a beamed-propulsion company, the Harvard Business School alumna and successful entrepreneur was already the youngest person to have completed neutral buoyancy training, fully suited underwater practice for space walks.

I won't go so far as saying I believe in fate, but it was an incredible coincidence that in a room of 1,500 dinner guests, I would be seated at the same table as Laetitia and Richard. For hours I bombarded them with questions about space exploration, and through their patient answers the enormous impact of the commercial space industry came into focus for me. By the end of the dinner, I had set a course on an exciting new professional goal. Behind every successful career you'll find a series of mentors and sponsors—mentors help point out the specific doors that you'll need to walk through on your professional journey and sponsors hold those doors open for you. Laetitia and Richard would prove to be both. They sensed in me a deep curiosity and an extreme work ethic, and over the years I aspired to live up to those expectations.

In the weeks after ECAD, I caught myself up on the past decade of advancement in human spaceflight. I learned that the end of the space shuttle program by no means signaled the end of NASA. In fact, thanks to the development of public-private partnerships, the 2010s were promising to be the most exciting era of human spaceflight yet. The door to the frontier wasn't closing, it was blowing wide open, and I had finally figured out what I wanted to do with my career. Over a series of follow-up coffees and lunches, Richard and Laetitia indulged another avalanche of questions about the burgeoning commercial spaceflight industry. When I expressed my intense desire to get involved and my willingness to contribute in any way I could, Laetitia suggested that Richard make an introduction to an organization they had long supported, a trade organization at the forefront of humanity's next giant leap. She explained the critical role of space policy and communication in the growth of a new industry. "If you're willing to go down to Washington," she said, "we'd be happy to introduce you to the folks at the Commercial Spaceflight Federation. They could always use an extra set of hands." Two days later, I was on a train.

Shortly before I showed up at his office, former astronaut and space station commander Michael López-Alegría had left NASA to take on the role of president of CSF. Mike was no stranger to commercial spaceflight; on his fourth and final spaceflight in 2006, his *Expedition 14* crew was joined by Anousheh Ansari, the world's first female private space explorer. Mike's esteemed NASA career made him a powerful advocate for commercial spaceflight, and I was amazed to find myself sitting across a desk from the astronaut who held the current American record for longest spaceflight and most extravehicular activities (EVAs).

I took a deep breath and laid out the pitch I had practiced on the train: I wasn't a technical expert, but I did have expertise in media and communications, and it seemed to me that one of the best ways to accelerate the progress of the commercial spaceflight industry would be to share more of the excitement with the public. The organization had a modest social media footprint, and I proposed a comprehensive campaign to grow its visibility and reach. If given the opportunity, I could help put into perspective the incredible societal impact of democratizing access to space and expanding Earth's economic sphere. That day, Mike L-A became the second person in the space industry (and second astronaut!) to take a chance on me. When I left his office that afternoon, I had a new role and a title that fulfilled the wildest of my *Star Trek* dreams: media specialist of the Commercial Spaceflight Federation.

By the time I joined the CSF team in 2012, the commercial space industry had long ago hit its stride, and a decade of historic accolades were already piled high across membership. I was determined to celebrate and communicate each and every accomplishment with social media posts and press releases.

While it was difficult to keep up with the volume of activity, it was even more challenging to keep coming up with synonyms for "historic." But how else to characterize major milestones where private companies were achieving what had only been managed by nations in the past?

Test flights were pushing the boundaries of vehicle design and performance and CSF members were making progress at a dizzying pace. I was a press release-drafting machine, celebrating an impressive range of milestones. This was the same year that SpaceX would become the first private company to berth with the ISS and the magnitude of that accomplishment blew me away. These weren't just historic achievements for an individual member company; they were historic achievements for the twenty-first century. CSF members were opening up access to space in a series of seminal moments not only for Americans but for the entire world.

The Commercial Spaceflight Federation Congratulates SpaceX on their Historic Achievement

MAY 25, 2012
Washington DC—The Commercial Spaceflight Federation congratulates SpaceX and NASA for the successful berthing of the *Dragon* spacecraft to the International Space Station today. NASA gave the green light for final approach after *Dragon* successfully demonstrated approach, pause, and abort maneuvers,

and other systems performance checks. SpaceX is the first private company in history to berth with the ISS.

SpaceX's *Dragon* capsule launched atop a *Falcon 9* rocket on May 22 from Cape Canaveral, Florida. In the days preceding its berthing with the station, *Dragon* has performed systems checks, maintained communications, and demonstrated various hold and retreat maneuvers away from the station and on approach. Once cleared by NASA, *Dragon* autonomously approached the space station, where it was grappled by the station's robotic arm and brought into berth.

CSF president and former ISS commander Michael López-Alegría said in a statement, "Tomorrow the six astronauts currently aboard the International Space Station will open the hatches to the first visiting vehicle from a private company. Future commercial cargo deliveries under NASA's COTS and CRS programs will ensure that the ISS continues to be a resource for America and its international partners.

"This is truly a momentous accomplishment for SpaceX and for the industry. The capabilities of the commercial space industry grow by the day, and America is well on its way to having a diverse, cost-effective, and dependable space transportation system. The entire team at SpaceX should be commended for their commitment and skill and thanked for their contribution to restoring US access to the space station."

By the end of my first year with CSF, my metamorphosis from curious space enthusiast to a passionate and persuasive advocate was complete. I subjugated myself to the mission: in addition to social media posts and press releases, I published editorials and gave countless public talks championing the success of the commercial spaceflight industry. I had absorbed a respectable understanding of the technical advancements and my media background enabled me to communicate them in plain terms. I tirelessly championed the promise of NASA's public-private partnerships—and the Commercial Crew Program specifically—to end American dependence on Russian infrastructure, ensure the success of the space station, and create thousands of high-tech jobs.

The ability to communicate is a powerful tool in an industry dependent on public support. The skill is helpful for increasing excitement and funding, but it becomes absolutely essential when you have to contextualize setbacks. That's a big part of why I was so insistent that we celebrate each member's milestone achievements, no matter how incremental they seemed. I wanted to drive home the reminder that space was a harsh and deadly environment, and that every hard-fought step to successfully operate a vehicle outside of Earth's protective atmosphere was accomplished in spite of this. I wanted to thoroughly memorialize the wins on the good days so that we could put into perspective the losses on the bad days. And as we all knew, losses would be inevitable.

Risk has always been intrinsic to spaceflight, part of the cost of opening new frontiers. As US President Ronald Reagan reminded schoolchildren in the aftermath of the 1986 *Challenger* disaster, losses are a painful but necessary aspect of exploration and discovery. "It's all part of taking a chance and expanding man's horizons. The future doesn't belong to the fainthearted; it belongs to the brave."

Failures are a certainty in opening a new frontier, and so for every upcoming test I would draft an outline of two different press releases: one celebrating success and another acknowledging that an anomaly had occurred. These so-called "good day/bad day" drafts have existed since the earliest days of spaceflight and have been on hand for some of humanity's greatest achievements.

The fact that *Apollo 11* moon landing was a historic first meant that the lunar lander had never fired its engines on the moon's surface before. When you add up all the unknowns and potential failures of a 1969 lunar landing, the risks seem incomprehensible: if that engine failed to ignite, Armstrong and Aldrin would have been stranded on the surface until their oxygen ran out; if the engine failed to burn long enough, they would have either crashed back to the Moon or been stranded beyond reach in low lunar orbit. Years of work went into de-risking the mission as much as possible, but there were still enough "ifs" to keep you up all night.

While Armstrong reportedly calculated the mission to be "90 percent safe," he understood there could be no guarantee. And he wasn't the only one. President Nixon's staff had prepared a "bad day" speech, one he would deliver in the event of a catastrophic failure. It was a speech that perfectly captured why the astronauts were willing to risk the odds and push the envelope in the spirit of exploration. Thankfully, Apollo 11 was a success, but had the lander experienced an anomaly, Nixon was prepared to address a grieving nation with the following remarks: "Fate has ordained that the men who went to the Moon to explore in peace will stay on the Moon to rest in peace. These brave men, Neil Armstrong and Edwin Aldrin, know that there is no hope for their recovery. But they also know that there is hope for mankind in their sacrifice."

During my time at CSF, successes vastly eclipsed setbacks, but when they did happen, we aimed to be as prepared as possible for crisis communications. That process was tragically tested on October 31, 2014, when Scaled Composites and Virgin Galactic encountered a tragic in-flight anomaly of *SpaceShipTwo* that resulted in the death of test pilot Michael Alsbury in Mojave, California. All flight tests by their very nature carry the risk of tragedy, but it doesn't soften the blow when it happens, especially when it hits so close to home.

The commercial spaceflight industry is powerful, but it's also tight knit, and every single one of our hearts was in Mojave that day. In an instant I was transported back to the day of the *Columbia* disaster, watching my television with disbelief as President George Bush comforted a shocked and grieving nation. I had learned a lot in the years since *Columbia*, and the echo of those words were so much more meaningful to me now that I was helping draft similar comforts for those who would look to CSF to understand the impact of this tragedy. I wished we had been able to write a different statement, one where Michael Alsbury returned to his wife and children after another milestone test flight that heralded the beginning of routine, commercial human spaceflight. But that wasn't to be. "Today, we are tragically reminded of the tremendous challenges that we face every day in our efforts to push the envelope of human experience and capability in space enterprise and exploration," began our statement. "The precious life that was lost cannot be replaced and will never be forgotten. The courage of both pilots will serve as inspiration for us all to continue to meet the challenges of spaceflight with clear focus and determination in order to make it as safe and reliable as possible."

The media aftermath was vicious and highlighted a capricious sentiment surrounding commercial space exploration. Just weeks before the accident, I had been in Mojave for the tenth-anniversary celebrations of the Ansari X Prize competition that started it all. Richard Branson had indicated that they were close to spaceflight, only to be met with sharp articles criticizing their incremental progress, with headlines quipping "It's Been Ten Years Since the X Prize—So Where Is My Space Taxi?" In the days after the tragedy, though, the narrative took a sharp turn, with inflammatory op-eds like "Enough with the Amateur-Hour Spaceflight," a scathing *Time* piece in which the broader industry was reduced to the tired trope of billionaire boys and their toys.

My heart was hurting for Mike Alsbury's family, for Virgin Galactic, and for the entirety of our industry. Articles like the one in *Time* were biting and seemed unfair. Virgin Galactic had uncovered a tragic engineering issue during the test, but didn't that speak to the very nature of a test-flight program? The explicit purpose was to trial technology and root out vulnerabilities before commercial operations. But now wasn't the time to correct the perception. If we wanted to count on the support of the public during the tough times, we were going to have to do a better job of bringing them into the fold during the great times, and that meant expanding our voice and presence. We were unified in the nobility of our purpose, but we had inadvertently cultivated an industry echo chamber, speaking most often to each other and others already "in the know"; I challenged myself to think creatively about broadening our audience and converting our fair-weather friends into fierce proponents who wanted to see this industry succeed because they, too, understood the costs and benefits of opening up new frontiers. We needed to do more than showcase objectively exciting flight achievements; we had to show our growing audience why test pilots like Michael Alsbury went to work every day in the first place.

In 2014, there existed a sentiment across the industry that "all boats rise with the tide." In other words, it was in everyone's best interest to support each other and be prepared for a show of resilience, because the activities of any individual company were often used to paint broad strokes across the entire industry.

Even minor setbacks would prompt a cluster of alarmists to wonder aloud, "Is this the end of commercial spaceflight?" Of course, that couldn't have been further from reality. Members of the CSF were democratizing access to space through incredible technology innovation and the progress was inspiring people of all backgrounds to rekindle their own personal dreams of space exploration. Those were the people I wanted to speak to directly, and my limited technical expertise helped ensure that I never lost the forest for the trees.

I started expanding our reach with a series of small forays into the world of modern media tools, cranking up our social media presence and coaxing Mike L-A to host "AMAs" on Reddit and Facebook Live. With the help of Richard and Laetitia Garriott de Cayeux, I convinced the entirety of CSF membership to join me at the Explorers Club for a public event we called "Blast Off: The Future of Spaceflight!" Representatives from SpaceX, Virgin Galactic, Masten Space Systems, Sierra Nevada, and a host of other private companies brought vehicle mockups and shared their progress with more than two hundred guests. I even convinced Blue Origin, our most notoriously private member, to participate. Not only did they show up, but they shared rare visibility into their flight test plans, which had space journalists buzzing. At that same event, we announced the creation of a new sister organization to CSF named Earth+, intended to raise awareness for the commercial space industry and encourage public involvement. As a non-engineer in a deeply technical industry, I had found my own ways to add value and advance the industry I loved.

By now, and thanks to the steadfast support and sponsorship from Richard and Laetitia, I was a member of The Explorers Club myself. As with any other door that was opened for me, I had committed 110 percent of my energy to the opportunity. In appreciation for the club whose membership I was so humbled to hold, I volunteered all of my spare time at the Upper East Side headquarters. No job was too small: during evening events I helped with the check-in and coat check, and when the front desk was short staffed, I was happy to reprise my old role as receptionist. My commitment to the club's most basic core operations earned me credibility for the more glamorous ones. That same year, I was asked to chair the famed Explorers Club Annual Dinner, becoming the youngest member in the club's history to do so. For this 110th ECAD, I had successfully lobbied for the theme of "Exploration & Technology," and along with the rest of the club's members I submitted my nominations for those I knew deserved the club's top honors. To my delight, the club's Flag and Honors Committee held my nominees in equally high esteem, and my worlds collided in the best possible way.

The Buzz Aldrin Quadrennial Space Award would be presented to former NASA astronaut and rocket engine innovator Franklin Chang-Diaz, as well as NASA geophysicist Maria Zuber. Buzz Aldrin himself would present the Citation of Merit to Blue Origin CEO Jeff Bezos and his Apollo F-1 expedition team, who had recovered the historic Apollo rocket engines from the bottom of the Atlantic Ocean. And my boss, CSF president and record-holding NASA astronaut Mike L-A, would present the prestigious President's Award for Exploration and Technology to SpaceX

CEO Elon Musk. Best of all, Professor Stephen Hawking had agreed to deliver a recorded keynote, and I was delighted for the opportunity to finally put my film degree to good use, traveling to Cambridge to produce the video and meet with him.

Chairing the event and sharing a stage with titans of industry was a humbling and powerful experience. I had produced an event that brought together some of the world's most brilliant minds to discuss their shared dreams of space exploration, and better yet, I had engineered potent moments of inspiration for 1,700 dinner attendees as well as hundreds of thousands more reached through media coverage of the event. I had found myself in the once unfathomable position of standing between Elon Musk and Mike L-A on stage, and from there I could see Laetitia and Richard Garriott de Cayeaux cheering me on from the same table where just two years earlier they had indulged my barrage of questions about how one could contribute to space exploration. I had given everything I had to ensure that their investment in me paid off, but it was a powerful reminder that cream doesn't always rise to the top on its own; sometimes it needs to be pulled.

ADVICE FOR ASPIRING
ASTRONAUTS

I'M NOT SURE I CAN PINPOINT A SPECIFIC MOMENT IN WHICH I CROSSED THE

threshold from space outsider to space insider, but it was around this time that I finally understood the difference between inclusion and belonging. I worked hard to break into the space industry, but that doesn't mean I did it alone. My career is full of people who gave me chances to prove myself. Sometimes those chances happen by, well, chance, but I don't believe in so many coincidences. Instead, I'm confident that there are certain behaviors that will better position you for those opportunities and make it easier for people to put their faith in you. If you're determined to take a more proactive approach toward your goals of breaking into a new industry, here's the advice I would offer:

1. Do Your Research, Get Involved

If you want to get involved in the action, you need to figure out where the action is happening. For the space industry, that usually means conferences. I've listed in the appendix some specific recommendations, but making yourself a regular fixture in your community's space scene is the surest way to build up a network. Many industry conferences will offer tickets in exchange for volunteer work, a role that often provides far more valuable access than attendance through general admission. Even if you don't have local event options or can't travel to a conference, you can still get creative. Join the conversation on social media, or if you're craving more direct contact, connect with your local university astronomy departments, planetariums, science museums, and libraries. Even better, start your own gatherings or digital engagement efforts. Spinning up a space trivia happy hour at your local bar or a space book club through your town's library can help establish you as a catalyst and connective force in your own community (and I have the perfect recommendation for your group's first read...).

2. Find Mentors, Earn Sponsors

The concept of a single, capital-M Mentor is antiquated and unrealistic, so if you're set on a rigid master-student dynamic, you're probably missing out on a whole lot of strategic guidance. Instead of a single specific person, consider mentors as a category of people from whom you can learn. If you've done step 1, you've likely amassed a couple relevant acquaintances, one of whom is likely to know someone who works in a field adjacent to your interests. I've always tried to ensure that the barrier to entry is as low and convenient as possible for anyone whose guidance I'm requesting; a coffee chat or quick phone call is a perfect opportunity to ask targeted, specific questions about their work and aim to gain perspective on how they would approach certain career decisions or obstacles.

The bar for sponsorship is rightfully higher. Everyone deserves some form of mentorship in their career, but sponsorship is earned when someone decides they're going to vouch for you, using their own reputation as collateral. I'm sure none of the folks in my life who have acted as sponsors viewed it as transactional, but my awareness and deep appreciation for their faith in me ensured I worked twice as hard to prove their investment was well placed. It wasn't enough for me to simply do a good job when people opened doors for me; it was important to me that I could repay the favor by absolutely crushing it and making *them* look good in the process. When someone puts their name on the line for you, you become a direct reflection of their judgment. But how to earn that initial sponsorship? First you have to develop a reputation of being worthy of it.

3. Design Your Ideal Reputation, Make It True

As an experiment, I asked a couple of people who have acted as mentors to describe me from their earliest impressions. I heard back "driven, passionate, and a ruthless executor." There are perhaps some deeper themes to unpack in that last one, but the overall point is that those descriptions aren't just happenstance; they're characteristics that I intentionally aimed to cultivate a reputation around. I didn't set out to appear "brilliant"; instead, I simply wanted to prove to people that I was mission-driven, action oriented, and reliable: you could count on me to get the job done, and I wouldn't rest until I did. I earned this reputation by tackling everything, even the smallest tasks, with 110 percent enthusiasm and attention to detail. My email responses were immediate, well-organized, and triple-checked for grammar and spelling. To make the most of people's time, I sent out agendas ahead of phone calls and meetings and ensured I was the first one on the conference line or at the coffee shop. I took notes and I closed the loop every single time; nothing died on the vine on my watch. If you recommended a book or an article, even in passing, I was going to read it, highlight passages, and report back to you. I sent thank you notes for people's time and I meant them sincerely. Your reputation is entirely within your control; decide what you want it to be (even just three or four adjectives) and make a plan about what behaviors you'll need to adopt or adjust to make those things true.

4. Seek the Important Work, Not the Glamorous Work

I've found that the most successful people don't chase their own success, they chase good outcomes. I want to be surrounded by people who are humble and gritty, hungry for results and willing to do whatever it takes to earn them. And 99 percent of the time, "doing whatever it takes" means you're doing the less glamorous work. While I've certainly relished moments in the spotlight (who doesn't?), I've developed the utmost appreciation for the work that happens behind the scenes, the kind that keeps the lights on. After years of drafting op-eds and press releases for others, I was thrilled to be invited to write pieces under my own name. But an exciting byline didn't mean I was suddenly above the rest of my responsibilities. Instead, I used my personal pieces as motivation to make my writing more impactful across the board. Even now, a published author in my own right, I continue to draft statements and communiqués for others because I care about outcomes, not credit. When you're truly focused on the outcome, *there is no job too small.*

5. Take a Breath, Get Your $%*^ Together

An organized life is a productive life. When you're juggling responsibilities, it can be a struggle to keep all the balls in the air. Leveraging productivity tools saved my time and my sanity. I relied on calendar reminders, daily knockout lists, and most importantly, I learned to live by what *Getting Things Done* author David Allen coined "the two-minute rule." The rule is this: if any task that lands on your plate over the course of the day can be accomplished in two minutes or less, you should tackle it right then and there. Instead of letting a to-do list of small tasks pile up, I actioned things in the moment, preserving my to-do list for things that truly required a block of time. This is especially important when the precious time you have to dedicate to your passion exists outside of your existing professional and personal obligations. Minutes add up to hours, and every hour counts.

6. Acknowledge Your Strengths and Weaknesses

It's worth investing some time in an honest audit of yourself. What are the types of problems you are uniquely well leveraged against? Maybe you're a great consensus builder, or storyteller, or creative iterator, or like me, a relentless executor who derives satisfaction from checking boxes. Take stock of those things that come naturally to you and aim to align yourself with as many of those types of tasks as possible. For me, this meant positioning myself in opportunities where I could provide operational rigor, letting my execution skills shine in the process. Similarly, take a hard look at your Achilles' heel. Those weaknesses aren't the same thing as the tasks you simply dislike (we all have grin-and-bear-it moments), but rather tasks where outcomes are at risk when you're deployed against them. Know the difference and avoid them at all cost.

7. Practice Your Personal Elevator Pitch

Inevitably, you'll find yourself in front of potential mentors, sponsors, and other folks with whom you're eager to leave a positive and lasting impression. To make the most of these opportunities, you'll want to have perfected your personal elevator pitch, a quick spiel about who you are and what you hope to accomplish. The best ones are succinct, confident, and unique. My personal pitch was also aspirational: "I'm passionate about expanding humanity's footprint in the solar system and currently serve as the media specialist for the Commercial Spaceflight Federation. I love representing our member companies, and eventually hope to fly on board one of those spacecrafts myself!" But before I broke into the space industry, it might have looked something more like this, "I'm a communications expert with a passion for human spaceflight. I work in the film industry and love bringing stories to life, but I've started to think about how I could apply that expertise to increasing awareness and support for commercial spaceflight. I'm currently looking for opportunities within the space industry." You'll likely find that your personal elevator pitch evolves over the years, so you make sure you revisit it annually for an update.

8. Re-Purpose Imposter Syndrome

The fact that some the most accomplished people I know suffer from chronic self-doubt convinced me not to dwell on my own feelings of inadequacy. I also learned that there's a big difference between being an imposter and an outsider. The differentiation was critical to growing my confidence while trying to break in to the space sector; to be an imposter would mean I was forging a career in an industry I had no right to be in, while being an outsider simply meant that I had work to do to earn my place. Women are especially vulnerable to this, particularly when it comes to applying for jobs. It's the difference between thinking, "I've got expertise in everything except x so I'm unqualified," instead of "I've got most of this and I'm confident I can pick up the rest on the job." If you wait until you are 100 percent confident in your abilities, you're sure to miss the opportunities that can actually help you reach expert status. This epiphany had become a long-running joke with Mike L-A, and when he eventually officiated my wedding years later, he kicked off the space-themed ceremony by giving loving props to the bride who had once "BS'd her way into the space industry."

9. Create a Personal Advisory Board

One of my most important acts of self-investment was the creation of a personal advisory board: an assembly of different people who cared about me and to whom I could turn for advice, encouragement, support, or perspective. It's important to have a group of people you trust for honest advice and opinions, and they don't necessarily need to be people in your field. Your personal advisory board can consist of colleagues, mentors, friends, relatives, and even mentees of your own. Well-established people often make the mistake of only looking up for inspiration, missing out on the creativity that can be shared from some of those folks at earlier points in their career. Once you've found your people, lean on them, and let them know they can count on you for the same.

10. Always Leave Your Comfort Zone

You likely have big dreams to accomplish; getting too comfortable can stall momentum and lead to stagnation. As clichéd as it may seem, you have abilities you might never unlock if you don't force yourself to try. Leaving your comfort zone means embracing that risk in the hopes of personal and professional growth. And for me, that eventually meant leaving the relative comfort of space communications for the new challenge of space hardware. By the end of 2014, I would be trading Washington, DC for Mojave, CA.

CHAPTER 5

TALES FROM A SPACEPORT

SPACE HAS BEEN A MILITARIZED DOMAIN

FROM THE EARLIEST DAYS OF THE SPACE

Race, lest anyone is under the illusion that Sputnik was a purely exploratory pursuit. It took me a few years to fully embrace a personal sense of responsibility for the defense side of "aerospace & defense," but exposure to the extent to which our country is powered by space-based technologies enhanced my perspective early on in my career. Just imagine a single day without space-based technology and the long list of modern conveniences that would be verboten on a space diet: no GPS, smart phones, electronic banking, satellite TV or radio. Also lost would be weather radar, storm tracking, and so many other Earth-imaging and communication dependencies; put simply, even a single day sacrifice of the satellites upon which our modern lives have become reliant could disrupt critical capabilities in the United States of America, and perhaps the entire global economy.

That's quite a vulnerability. More alarming, the above was imagined as a voluntary sacrifice, pre-determined and timeboxed. Now imagine the impact of an unplanned threat to our space-based assets, whether through hostile takedown or accidental debris impact. The further we advance our space technology capabilities, the greater the extent that human lives depend on them, cementing space as a major national security issue. Like air, sea, land, and cyberspace, space has long been a domain worthy of equal attention from the United States military. In fact, "Day Without Space" exercises have already been war-gamed in the Air Force through a series of military drills predicated on a hypothetical total loss of space capabilities.

From the earliest government rocket tests, these US military interests have underpinned space activity and protected American lives, assets, and interests, and have enabled rapid response to a number of global natural disasters and humanitarian crises. Militarization is necessary and well established, but perhaps a timelier societal conversation is the *weaponization* of space. The 1967 Outer Space Treaty prevents weapons of mass destruction in orbit or on other celestial bodies, and while that seems like a wise piece of governance, modern technology in the half century since Apollo has surfaced plenty of potential grey areas. Hypothetical loopholes (what about nuclear weapons not technically in orbit, and exactly how much destruction is "mass" destruction?) are enough to keep anyone up at night, and provide reasonable justification for militaries (and Space Forces!) around the world to continue fortifying their own space-based operations and defense capabilities while simultaneously attempting to define and regulate the offensive use of those space systems.

US president Dwight D. Eisenhower foresaw the inevitable bifurcation of military and civilian interests. In 1958, he authorized a two-pronged approach to advancing America's spacefaring capabilities and resiliency. In the Department of Defense, the Advanced Research Project Agency (ARPA, and later, DARPA) was established and tasked with overseeing all military space activity. The same year, Eisenhower and Congress signed the National Aeronautics and Space Act to establish NASA, the beloved and much more visible agency whose mission served the civilian half of space technology and exploration.

Given the advancements of the commercial spaceflight industry, it shouldn't have surprised me to discover NASA wasn't the only American agency looking to the private sector to complement their capabilities in space. DARPA, sometimes referred to as the DoD's mad scientist lab, was also interested in leveraging private industry for American defense and national security goals. DARPA saw the benefit of reusable rocketry for short-notice spaceflight, where payloads could be launched from anywhere in the world in a matter of days, rather than months.

To achieve this goal of low-cost, routine access to space, DARPA envisioned an un-crewed, reusable spaceplane that could carry payloads to low Earth orbit on extremely short notice. This would create a significant defense advantage, providing contingency plans for those military and commercial satellites on which American safety, security, and prosperity are most critically dependent. Should one of those satellites suffer a catastrophic loss, DARPA envisioned it could be rapidly restored by this experimental spaceplane (XS-1), and they were eyeing private industry to build that vehicle.

To prove out the technology, DARPA settled on an initial list of criteria that private industry partners would need to achieve in a demonstration flight: the XS-1 would need to fly ten times in ten days, hitting speeds of Mach 10 or higher, while carrying payloads between three and five thousand pounds in weight. And it would all need to be done for a cost of less than five million dollars per flight.

In addition to my role at CSF, I was an active member of the Space Frontier Foundation, a space advocacy group long known for promoting collaboration between government and private industry. When DARPA wanted insight into how best to position the program for commercial sustainability, they knew exactly who could help conduct a study. As with many public-private partnerships, the XS-1 program was structured in a way that encouraged the private sector to create a business case for the rapid-launch services they would enable. And it was in support of this goal that I partnered with DARPA's Tactical Technology Office to hold a series of industry workshops to gather data and feedback on issues related to the transition of XS-1 technology into the commercial space industry.

Both industry and government would benefit from rapid and affordable launch capability; cheap access to space meant lower barriers to entry for commercial payloads, which carried the potential for a new market of novel broadband systems and other innovations—a huge incentive for vehicle developers. I assembled an impressive group of industry leaders, and through a series of workshops, we unpacked the technical, budgetary, and programmatic elements of the XS-1 program in excruciating detail, aiming to uncover as many potential commercial and customer applications as possible.

I was earning a reputation as a powerful connective force in the space industry, stretching my skill sets and my perspectives in the process. I had come a long way from drafting press releases alone, but I still loved writing statements that celebrated the achievements of CSF's member companies. And in July of

2014, when DARPA awarded phase one study contracts to three companies, I drafted a statement that changed the course of my career and presented an exciting new challenge.

CSF Congratulates Masten Space Systems on their Partnership with DARPA

Washington, DC—The Commercial Spaceflight Federation congratulates executive member Masten Space Systems for its selection to partner with the Defense Advanced Research Projects Agency (DARPA) as part of the Experimental Spaceplane (XS-1) program. Masten will use their expertise for the development of a reusable launch vehicle capable of flying ten times in ten days and lifting payloads greater than three thousand pounds to low Earth orbit.

"The Masten team has had incredible success in developing reusable vertical takeoff/vertical landing vehicles," said CSF president Michael López-Alegría. "With multiple flights already flown under the NASA Flight Opportunities Program, their experience will be advantageous in the development of the XS-1 vehicle. I look forward to watching their progress as they work to create safe, reliable, more routine access to space."

DARPA awarded phase one development contracts to three companies: Boeing, Northrop Grumman, and Masten Space Systems. Competition would be fierce, and from the perspective of a small team in the desert, this was shaping up to be a battle along the proportions of David vs. Goliath. Or more precisely, David Masten vs. Goliath.

Like so many space industry leaders, Dave Masten was a bona fide rocket maverick. Unlike others in his commercial space industry cohort though, he eschewed the spotlight, rarely granting interviews and deliberately claiming the title of CTO, rather than CEO, of his namesake company. Oh, and he also didn't have a billion dollars. What he had instead was a long-sustained passion for space exploration, a penchant for rocket engines developed over years of hobbyist tinkering, and a successful career in IT networking and software. The success led him to found Masten Space Systems in 2004, where he could fully invest in his dreams of lowering the barriers of access to space. Focused on the enormous and costly delta between maintenance and servicing for a space shuttle versus that for a commercial airliner, Dave tackled reusable spaceflight with an eye toward lean operability. Headquartered in the space industry's version of Silicon Valley, a modest workshop at the Mojave Air & Space Port proved the perfect home for reusable rocket-powered vehicles that were designed to be operated by a small team, and just in time for a brand-new X Prize.

In 2009, NASA was interested in paving the way for the next generation of lunar vehicles. While rockets typically go straight up, the NASA and Northrop Grumman Lunar Lander Challenge required competing teams to launch, fly laterally, hover in place, and then land their rockets with precision on another pad. This was the perfect challenge for Masten, whose VTVL rockets had been designed with guidance, navigation, and control capabilities that the team hoped would allow the vehicles to pirouette atop a narrow plume of fire.

Sure enough, through short, low-altitude rocket-powered hops around the Mojave Air & Space Port, Masten's XA-0.1B ("Xombie") rocket launched, hovered, and landed just inches from its target, becoming the first vehicle of its kind to take to the skies at Mojave Air & Space Port. For the one-million-dollar X Prize purse though, Masten's second rocket, Xoie, would need to do the same thing on a flight path that now included boulders and craters to mimic the lunar surface and test the vehicle's hazard avoidance capabilities. What followed is a Masten Space Systems origin story passed down to every new hire, a tale of stress and triumph that perfectly captures what it means to be a part of a team willing to give it their all.

Competition day started out with a series of problems. Problem one: Xoie was having trouble starting. Problem two, and more concerning than problem one: Xoie was leaking propellant. To force an engine ignition and compete for the one-million-dollar prize meant risking the $300,000 vehicle, and failure would likely bankrupt the start-up. In an ethos-defining move, the team unanimously decided they were all in. Xoie's engines finally ignited and she flew the entire course, only to have her oxygen tank burst into flames upon landing.

Luckily for the team, the judges showed mercy and granted Masten one final flight attempt at sunrise the next morning. This left the already sleep-deprived team less than twelve hours to find the leak and repair the charred rocket. In what can only be described as a Mojave miracle, members of other competing teams swarmed in to help rebuild the rocket overnight.

The next morning Xoie rose from the ashes, the leak solved with a Rubbermaid trash can lid and some baling wire. The rocket executed a flawless flight and precision landing that clinched Masten the one-million-dollar prize purse and demonstrated that commercial industry could indeed be leveraged to advance the entry, descent, and landing (EDL) technologies necessary to land safely on other celestial bodies. The win cemented Masten's reputation as a leader in rocket-powered launch and landing and the scrappy start-up proved to be the spaceport's most prolific tenant. In a market where rockets only went up, Xombie, Xoie, and a small army of X-named successors performed hundreds of acrobatic flights for customers who needed to simulate a custom flight path or who wanted to test and mature their payloads in true rocket-powered conditions.

In 2010, Xombie pushed the envelope even further. The rocket launched and paused in midair while the engine purposefully cut off, before reigniting and performing a nail-biting in-air engine relight, becoming the first VTVL vehicle to do so. The dramatic video footage generated positive attention from both NASA and the industry. It even reached Elon Musk, who shared the video link with his propulsion, avionics, and structures teams at SpaceX. Masten still flew under the radar, but the company had earned recognition as a reliable platform to prove out game-changing technology.

I loved my work with CSF, but I knew I hadn't yet found my ceiling as a non-engineer in the space industry. I was already starting to search for my next challenge, ideally working for one of our member companies. I had initially been looking at bigger names, but DARPA's XS-1 program had put Masten Space Systems on my radar. To some companies, a three-million-dollar first phase award was a drop in the bucket, but to a team of fewer than twenty employees in a small Mojave workshop, three million dollars was transformative. DARPA knew exactly how far that investment could stretch, especially from a team who had repeatedly proven that they were willing to give everything they had.

The small but talented team had earned a reputation for being humble, scrappy, and ruthlessly efficient. There was no billionaire backing here; by necessity, Masten was capable of doing more with less, and that resourcefulness was baked into their DNA. More than that, it kept the lights on over the course of a decade when others in Mojave were shutting down. Masten had been demonstrating rapidly reusable rocket-powered flight for years prior to the XS-1 award, and their small size provided operational advantages for a program that relied on small crews for fast turnarounds at low launch costs. By 2014, a number of small companies and fellow X Prize competitors had sadly gone the way of Rotary Rocket, but Masten Space Systems was both surviving and thriving, now with an opportunity to scale their work.

I had already established a working relationship with Masten CEO Sean Mahoney through CSF, and when I heard he was looking to expand their team of creators, builders, and doers to enable both exploration and defense, I knew immediately that this was the mission I wanted to join. On my first visit to the company's humble headquarters in Mojave, Sean ended the informal job interview with a pointed stroll past Rotary Rocket's permanently grounded test vehicle, which he described to me as a monument to failure. "There's no middle ground here," he said. "This is what it looks like to go all in." And in the desert heat, I signed a job offer that same afternoon.

Part of Mojave's charm lies in the contradiction of innovation and inertia. Pulling off the blink-and-you-miss-it highway exit eighty miles north of Los Angeles, you'd never suspect that some of humanity's boldest achievements were being engineered in the dusty railroad town. There, past cement blocks and tumbleweeds, a weathered billboard marks the humble entrance to the Mojave Air & Space Port, summing up nearly one hundred years of aviation ingenuity with the understatement of the century: "Imagination Flies Here."

You can trace Mojave's "dream big" attitude all the way back to the earliest days of the airport, when two dirt runways opened up to serve the 1930's gold mining industry. The rural desert was blooming with activity and optimism, and by 1941, the town was eyeing updates to the modest airfield. The US government agreed to sponsor the improvements with one prescient caveat: the airport could be commandeered by the military in the event of war. The ink had barely dried on the contract when the attack on Pearl Harbor and the resulting entry of the United States into World War II activated that very clause. The United States Marine Corps transformed the Mojave Airport into an auxiliary air station, including facilities and barracks to house more than three thousand squadron members supporting the war effort.

Over the years, the stretch of restricted airspace over Mojave and nearby Edwards Air Force Base would provide a backdrop to a number of aviation "firsts"—in 1947, Chuck Yeager become the first pilot in history to exceed the speed of sound in level flight, kicking off an impressive succession of sound-barrier-breaking flights above the Mojave Desert. A few decades later, those sonic booms would announce the return to Earth of NASA's first space shuttle, followed twenty years later by the supersonic flight of *SpaceShipOne*. By the time the commercial space industry moved in, Mojave had already hosted a number of adrenaline-pumping predecessors.

It wasn't until the 1960s that the local county reclaimed the deed to the airport from the military, and in the decades leading up to the Ansari X Prize, a unique subset of aviation activity found a home at the Mojave Airport. A rich history of air racing, flight testing, jetliner storage and scrapping, and even the filming of iconic Hollywood stunt and car chase scenes made Mojave the natural choice for a new space industry looking for permission to push the envelope.

The technology was advancing rapidly, but everything else in Mojave stayed more or less the same. Many of the World War II facilities remained standing, the barracks converted into commercial warehouse storage and workshops. Just days before *SpaceShipOne*'s history-making flight in 2004, Mojave Airport was redesignated the Mojave Air & Space Port, hanging a shingle for companies like Masten who were eager to set up shop and push the boundaries of rocket science.

"Imagination Flies Here" was more than a slogan; it was equal parts heritage and challenge. It was also the embodiment of former Mojave CEO Stu Witt's attitude toward private spaceflight. "My job is to give permission," he explained on my first tour of the spaceport. "Every day in the skies above us and on the ground here at the spaceport, people are taking incredible risks for the progress of humanity."

And permission was an important distinction—the California desert certainly offered companies an optimal landscape for flight, but it was Stu Witt's leadership that enabled them to take those risks in the air. Throughout more than a decade of leadership, Stu was a fierce champion of the commercial spaceflight industry, even serving as chairman of the CSF. He was an aerospace optimist, but he was also as a realist. Permission to succeed comes with permission to fail, and Stu would be the first to acknowledge that permission can come at a cost. In Mojave, that cost stood on display as a sobering collection of plaques in the Legacy Park memorial garden, a tribute to those who had lost their lives at the spaceport in the pursuit of opening up the final frontier. An inscription of "*Ad astra per aspera*," summed it up best, Latin for "Through great hardships to the stars," and a somber reminder that frontiers favor the brave-hearted. Giant leaps forward require bold vision and leadership. With Stu at the helm, imagination was permitted to fly in Mojave.

Here's the thing: no one moves to Mojave to take a job. The work is hard, the conditions are unforgiving, and the salaries are a far cry from nearby Silicon Valley. You move to Mojave to take part in the action, to subjugate yourself to a cause bigger than yourself, and to earn some space stripes along the way. Like the early entrepreneurs who moved to Mojave to help open up gold mining operations, and those who followed in support of the war effort, I too made my way to Mojave to help secure the promise of a brighter, more prosperous future. This time, through space travel.

It was a career move that exposed the limits of personal advisory boards, and it had mine divided. Half recognized the potential of Mojave as an enormous growth opportunity, both personally and professionally, but the rest of my well-intentioned advisors suggested I consider more established companies, where I wouldn't necessarily need to join at the ground floor. For my next career step, though, the ground floor was precisely what I was looking for, or more specifically, a workshop floor, where I could contribute to space exploration in a more tactical, hands-on way than before. I wanted to join a small team where I had the opportunity to earn a number of different responsibilities and create an outsized impact.

As a non-engineer at a larger space company, I likely would have slotted into an established communications or business development team within the broader organization. Chances are high that I would have found myself in a glossy office far away from the hardware and operations. And while I'm certain I would have loved every minute of it, I know now that I would have missed out on an enormous opportunity to stretch my skill sets and probe their boundaries. Rather than gloss, I was looking to get my hands dirty. And that dirt was pretty much guaranteed in the Mojave Desert. The spaceport and the Mariah Inn (now Comfort Inn) next door, became my home away from home, a major commuting convenience for a job where the workdays kicked off shortly after dawn.

Like many other companies in Mojave, the official headquarters of Masten Space Systems was a military relic from the 1940s, one of the many barracks converted into office and warehouse space. My official title was "Business Development Specialist," but as is required on a small team, and exactly as I was hoping to do, I had the privilege of wearing many different hats. In fact, I have never before (or since) held a job where I spent less time behind a desk.

Masten's phase one contract for DARPA's XS-1 program was a big deal and hosting top military brass in our humble barracks-cum-workshop was a thrill. But beyond those programmatic review meetings, plenty of other work ensured my hands and mind were never idle. I spent hours in our rocket garage, working side by side with the NASA scientists and academic research teams to supervise the integration of scientific payloads onto our rocket-powered landers, which would serve as the trusted test beds for the guidance, navigation, and control systems necessary to land spacecraft safely on the surface of the Moon or Mars.

The vehicle garage was exciting, but our test site was exhilarating. A short drive past warning signs and a badge-activated gate earned you entrance to the launch pads, where we slowly and lovingly towed our vehicles out to our dedicated slab of concrete before settling into the matching bunker a few hundred feet away. There, we put to use the sunrise safety briefings that kicked off each test morning.

Safety would rightly be a recurring theme during my time in the desert, but close-toed work boots and dressing in layers protected you from a whole lot more than engineering injuries. The Mojave Desert is the driest desert in North America, and while the flat and sparsely populated land offered an optimal rocket testing environment, the desert was home to plenty of other species. During my first week in Mojave, I observed scenes from both *Planet Earth* and *Arachnophobia*. The most traumatizing memory blends the two, in which I watched North America's largest scorpion (literally called the "giant hairy scorpion") battle a black widow spider a few feet away from where I was standing. I didn't stick around long enough to see who won, but I was an unwilling spectator to dozens of other venomous fight clubs during my time in Mojave. In each instance, Dave Masten offered the same unbothered words of wisdom: "If you don't interfere with them, they won't interfere with you."

The desert environment also required some adaptation. Temperatures bounced between an extreme binary of blistering heat and frigid cold; I could deal with either one, but it took some adjustment to get used to experiencing both within the span of a single day. Sunscreen, hats, layers, and hydration were necessities in an environment that could deliver both heatstroke and hypothermia on a whim. And then there were the actual rocket-related hazards that had brought so many of us out to the desert in the first place, loading combustible fuel into rocket engines that, if ignited correctly, promised eventual sonic booms that would split the air and rattle the building frames.

Masten's explicit mission was to engineer a future where rocket flights are such a regular, routine occurrence that they become boring. While I championed the vision and sentiment, I knew from the very beginning that there would never come a day when the novelty of rocket-powered flight was lost on me. By the time my head hit the pillow each night, my thoughts were consumed with the same sense of amazement: I couldn't believe I was working with rockets at an honest-to-god spaceport.

At Masten, everyone was expected to contribute in every area. I wasn't an engineer, but I shared with my colleagues an intense mission focus and a bias toward action. In that sense I was able to default to a familiar operating zone, where no job was too small and no challenge too daunting. I didn't need to be the smartest person in the room, but I always aimed to be among the hardest working. I dedicated myself to diagnosing the business priorities and developing the skills necessary to help execute against them.

We were launching rockets on a relative shoestring budget, but Sean and Dave invested everything they had to help me ramp up on rocket science and build operational muscle in the areas in which I had both interest and conviction. When I expressed a desire to better understand how exactly the flow of cooled isopropyl alcohol and liquid oxygen powered our rocket engines, Sean had me in a hazardous materials handling course at dawn the next day. Instead of hearing or reading about how the propellants worked (or worse, being discouraged from veering too far from my lane), I could get certified to help load them into the rocket myself and watch the engine ignition sequence take place on the pad. Through hands-on learning, Sean helped me convert unbridled enthusiasm into a focused and nuanced passion for space exploration.

That level of individual investment wasn't simply luck or good leadership; it was a carefully fostered cultural gem. The key to a team capable of doing more with less lies in a flat organizational structure where everyone feels equal ownership of outcomes and an equal responsibility to invest in each other to reach them. By morning I might be sitting side by side with our flight controller in the bunker getting trained up on telemetry readings, and by early afternoon I would be fastening down a helmet visor while another colleague coached me through a lesson in the surprisingly delicate art of TIG welding. It took a long time for me to be able to lay consistent metal beads, but to this day, there's a rocket-powered airframe flying around Mojave that I can say I helped build.

I worked hard to build credibility with the engineering team, but there were also plenty of areas where I could lean on my own strengths. My eye for storytelling and ability to translate our technical mission for the public led to a more robust media presence for the company. I hit the road with Sean to take the stage at conferences, panels, keynotes, and other appearances that provided an opportunity to share our story and help government, academia, and industry imagine the ways in which they could use our platforms to mature their own space technologies.

With a little bit of coaxing, I pushed Sean and Dave toward in-depth interviews with major publications and helped prep the team for those opportunities to showcase our work. I wanted people to think of Mojave and companies like Masten when they thought of blockbusters like *Star Wars*. It was important to me that people understand that those dreams of planetary settlement and cosmic expansion aren't limited to science fiction; the foundation for those bold leaps was being laid right here in Mojave and companies like Masten were actively engineering that future.

That everyone at Masten had once looked to *Star Wars* for inspiration was expected, but it was incredible to discover that *Star Wars* also looked to Masten. One of my favorite days in the desert was a visit I organized from George Lucas's Skywalker Sound team, who joined us behind the bunker to record rocket launch audio that would be mixed into the sound effects for *Episode VII: The Force Awakens*. As a special thank you, the Skywalker team created for us a short, custom clip of our launch audio remixed into an imagined Tie Fighter engine test, complete with audio dialogue between two pilots (non-canon, but no complaints!) Spending the day with sound legend Ben Burtt and the Skywalker Ranch crew drove home exactly what all of us were doing in Mojave every day: we were bringing science fiction to life.

From the moment I understood the value and responsibility of advancing humanity's footprint in the solar system, I had dedicated myself both personally and professionally toward that goal. At Masten, all cylinders were firing on the professional side; each workday was spent maturing the technologies that would open up the unexplored corners of our solar system, but I still wondered whether there was more I could do personally. I had passion and aptitude, qualities with which I hoped to continue contributing in the quest to move beyond flag-and-footprint missions, and instead toward long-term, sustainable human presence on another celestial body. I wanted to see boots on Mars in my lifetime, all the better if they were my own, but I knew we had a long way to go to mature the technologies that would enable that next giant leap. Luckily, that research was happening right here on Earth, and once again I found myself with an opportunity to get my hands dirty and contribute directly. With the support of Masten, I was soon packing my bags for a crew rotation at the Mars Desert Research Station.[1]

....

1   DARPA eventually selected Boeing for Phase 2/3 of the XS-1 Program in 2017, including $146 million in DARPA funding along with an unspecified investment from Boeing. By 2020, Boeing had pulled out and effectively ended the program. Masten continues to invest in rocket-powered landing technology and is currently focused on supporting NASA's return to the Moon.

CHAPTER 6

THERE WILL BE BEER ON MARS

FROM THE TIME I WAS A CHILD, SKYDIVING
ALWAYS HOVERED NEAR THE TOP OF MY

bucket list. In the years since crossing that experience off, I find it funny that the most vivid memory from my first time wasn't the exhilaration of free-falling through the clouds, but rather the exasperation of the friend I had tried (and failed) to coax into joining me. She was resolute in her resistance and baffled by the appeal: why on earth would anyone want to jump out of a perfectly good airplane? Fair enough. It occurred to me that there are people who find the idea exhilarating, people who find it horrifying, and very few in between.

I've noticed a similar binary with the topic of space settlement. For some, the allure is obvious. For others, it's difficult to understand why anyone would want to leave a perfectly good planet. After all, we've got the best beaches in the solar system! Keeping both feet planted on terra firma seems like a reasonable instinct, but just as a parachute starts to sound a lot more appealing when imagining a doomed aircraft, so too will the allure of space settlement increase along with societal awareness of an entire species aboard a doomed planet. To remain on Earth without developing the capabilities to spread out into space means extinction. And we don't want to wait until the planet is on fire to figure out how to advance our footprint in the solar system.

I once had the opportunity to pose this question of urgency to Professor Stephen Hawking, asking him to share his beliefs on the importance of space exploration and settlement as part of his keynote address for the Explorers Club Annual Dinner. Through careful, deliberate word selections transmitted through his speech synthesizer, the world-famous physicist was all too happy to share his perspective. "Not to leave planet Earth would be like castaways on a desert island not trying to escape," he warned. "Sending humans to other planets will shape the future of the human race in ways we don't yet understand, and may determine whether we have any future at all," he said, before wryly adding, "...and if there is an asteroid on a collision course with Earth, not even Bruce Willis could save us."

This isn't to say Professor Hawking was pessimistic about the future of our species; far from it, he is someone who spent decades translating the beauty and wonder of the cosmos. Rather he was one of many voices pointing out the scientific fact, however grim it might seem, that at some point, planet Earth will cease to support life. It may be an incredibly long way off, or it could be much sooner, but one thing is certain: there is no future without space settlement. As Professor Hawking noted, a single asteroid could wipe out all of humanity. And even if we're fortunate enough to avoid that fate, he explained, the end of all plant and animal life on Earth will come about a few billion years from now, as our dying sun exhausts the last of its nuclear fuel, violently expanding into a red giant that envelops our planet. "Trust me, we don't want to be around when that happens," he used to joke.

The timeline may seem impossibly far, but it's quite likely things will go downhill much sooner. Temperatures on Earth will continue to rise, and some of the earliest victims may be plants, who paradoxically start absorbing less carbon dioxide when they sense their moisture draining, effectively restricting their photosynthesis fuel and accelerating our carbon emissions. When plant species are threatened, so too are the animals and humans that rely on them for food and oxygen.

That was all still related to the long-term expiration of the sun—we hadn't yet discussed the things that could happen in the near-term, over the next few centuries. Because he's super fun at parties, Professor Hawking reminded me that, beyond total ecological collapse and nuclear war, infectious disease was another area of grave concern for the species. He didn't live to see 2020, when the world had a small taste of how quickly and completely a global pandemic can bring an entire species to its knees, but I imagine he wouldn't have been entirely surprised. I like to think he would have listened to my (very animated) account of the global situation before asking, "And how's our space program coming along?"

Besides events (somewhat) out of our control, it's a troubling irony that the dawn of a Space Age coincides with a higher likelihood than ever before of wiping ourselves out—all the more reason to take advantage of this unique window in human history, the first time in more than 4.5 billion years that interplanetary life is possible. Beyond mere survival, there's also the prosperity and progress of our species to consider. Through space exploration, we aim to raise the global quality of life on Earth through innovation, knowledge, technological advancement, resource replenishment, and new economic opportunity.

Through my work in the commercial spaceflight industry, I understand that space settlement is more of an economic challenge than an engineering challenge. An abundance of resources and rhetoric had propelled a nation to the Moon; with similar funding and public support, the engineering capability exists to make life multi-planetary. The challenges to secure the national commitment, public support, funding, and even legal clarity are enormous, but there are plenty of folks willing to try.

Space settlement is an ambition that has captivated a number of space programs, companies, individuals, and nongovernment organizations. Some of the longest and most influential advocacy efforts have been driven by the Mars Society, a worldwide nonprofit organization dedicated to promoting the human exploration and settlement of our next-door neighbor. The organization has made a number of compelling arguments for establishing the first permanent human presence on Mars, as opposed to the Moon or an orbiting space settlement.

Mars Society founder Dr. Robert Zubrin has long argued that the red planet is rich in raw materials and energy sources that we could harness to support life. In addition to nitrogen and carbon dioxide gases, Mars has ice and the potential for liquid water. Perhaps most attractive is the atmosphere on Mars; while thin, it provides at least some minor protection from solar flares and other cosmic hazards while still absorbing enough light to allow solar power as a viable energy source for a space settlement.

On the flip side, the environment of Mars, like the rest of space, is downright hostile. While human spaceflights in low Earth orbit have surpassed one consecutive year, the longest time ever spent *beyond* low Earth orbit, traveling through the Van Allen radiation belts, remains just twelve days with Apollo 17. Put into perspective, robotic missions to Mars have taken anywhere from four to ten months to reach the red planet, a journey rife with dangerous radiation exposure.

Innovation from the commercial spaceflight industry could help dramatically reduce the travel time, but even once you've safely arrived, radiation on Mars remains a critical concern. If that's not enough, there's plenty more to worry about: temperatures can drop ferociously low and the extremely thin, carbon-dioxide-filled atmosphere would kill you if exposed. To survive, humans would need complex life-support and resource-processing systems, ideally sourcing and recycling locally found materials like water and soil. That said, the temperature and sunlight conditions on Mars are closer to Earth's than anywhere else in our solar system, (except perhaps for the cloud tops of Venus, which would make for a significantly more complicated engineering challenge).

I was location agnostic when I first approached space settlement research; I was simply motivated to see it happen. I recognized the great fortune to be alive in the unique window in history where interplanetary life was possible. Due to the existential threats previously described, that window won't be open forever, and we can't afford to wait until the very end to take action. While the opportunity is in front of us, right here and right now in the Space Age, we have to live up to our era and begin laying the groundwork for humanity's astronomical trajectory. Our next giant leap will see human civilization step boldly into the solar system, but the first small steps are taking place right here on Earth, in analog environments designed to mimic the challenges of life in space.

Traveling to space is difficult. Living in space is even harder. Whether in orbit, on the Moon, or on Mars, early settlers will need to be physically, mentally, and emotionally prepared for the daily obstacles of life off Earth. Luckily, our home planet is rich with geological diversity, providing plenty of terrestrial analog environments where astronauts can test the equipment, procedures, and temperaments they'll need in space.

The use of these so-called analog environments isn't new. Before Apollo astronauts traveled to the Moon, they spent weeks traipsing through a small fishing village in northern Iceland, where they practiced collecting and photographing rock samples from the basaltic, moon-like landscape. This field training was a tactical complement to the academic instruction, allowing them to practice and perfect the physical procedures required to complete the scientific objectives of their mission. In addition to Iceland, training brought Apollo astronauts to a number of other sites that mimicked the barren lunar landscape, including Arizona's Cinder Lake Crater Field, the Grand Canyon, and even the dramatic volcanic ridges of Hawaii.

Of course, the Apollo missions were designed as roundtrip flights, so the analog preparation was limited to geological field training, ensuring astronauts could make the most of limited time on the lunar surface. In recent years, though, as NASA and private industry have contemplated more permanent human presence in space, analog research has evolved to address a much broader range of training. In remote corners of Hawaii, Utah, Texas, Antarctica, and even in underwater facilities, research crews have begun sequestering themselves in closed habitats for weeks, months, or years at time, embarking on simulated missions designed to mimic even the most mundane aspects of space settlement and long-duration space travel.

The Mars Society owns and operates the Mars Desert Research Station (MDRS), a prototype laboratory tucked away deep inside Utah's San Rafael Swell, where a millennia of iron oxide dust has stained red the ruggedly beautiful rock formations. The dramatic landscape bears stylistic and geological similarities to Mars, and since 2001, a number of national space agencies and scientists have utilized the remote facility to conduct analog Martian field research. The habitat offers a simulation of life on the red planet, and the prototype laboratory is a serious scientific base for researchers dedicated to advancing humanity's ability to live off Earth.

In addition to the main Hab (habitat), the MDRS boasts two equipment-packed observatories, a climate-controlled GreenHab that houses aquaponic and conventional growth systems, an ATV/rover repair and maintenance module fashioned out of an old Chinook helicopter, and a geodesic-dome-turned-scientific laboratory.

The existence of a Hab on Mars presupposes a few steps on the path to space settlement: first, the construction of a move-in-ready facility would likely require a number of robotic precursor missions, delivering the equipment and supplies to the settlement site. We would need some form of robotic assembly as well; you can even imagine a rover constructing landing pads for those initial cargo missions from the Martian raw materials. Among the equipment ferried from Earth, you'd find medicine, tools, nonperishable foods, and life-support systems. Self-sustainment will require advancements in technology to convert renewable Martian raw materials (nitrogen, carbon dioxide, dirt, ice) into usable resources (water, fertilizer, methane, and oxygen).

Along with air and food, water is a big one. The average person consumes about half a gallon of water per day to stay hydrated, but we're accustomed to using nearly one hundred gallons each day over the course of standard activities like bathing, laundry, dish washing, and toilet flushing. That level of consumption is simply not an option on Mars. Resupplied resources from Earth would be a rare and expensive endeavor, so early settlers would need to recycle water and oxygen (or convert them from whatever natural forms exist on Mars) and treat them like the precious resources they are.

Early settlements will require some basic utilities as well. The Martian environment is a deadly one, and systems will need to be fortified against that hostile environment, ideally minimizing the need for crew members to suit up and journey outside to make repairs or assess damage. Solar power will generate electricity,

THERE WILL BE BEER ON MARS

but crews will still require energy storage deposits for when violent dust storms block out the sun's light.

Communications are also sure to be an absolute boondoggle for the earliest explorers. From Mars, the speed of light is a limiting factor, and depending on the orbital position, one-way communications between Mars and Earth can take anywhere from three to twenty-two minutes. Perhaps future satellite constellations can solve for this obstacle, but until then a connection to home is unreliable at best. Real-time communications like telephone calls or video conferencing are downright impossible with current technology.

All that is to say you're pretty much on your own, so it would be in a crew's best interest to have a team of cross-trained generalists capable of solving any problem that might arise, whether medical or engineering shaped. While robotic precursor missions can certainly help establish initial colonies, a human settlement mission to Mars must be approached with the experiences of all other frontier-opening endeavors in mind: people will die on Mars, some of them naturally and others directly in pursuit of opening that frontier.

Perhaps the biggest necessary adaptations will be social and psychological, as astronauts will be cut off from their home planet and the rest of their species, and deprived of familiar earthly comforts like nature, weather, and fresh air. It's a situation that many would find unappealing, if not downright intolerable, but thankfully there are plenty of bold explorers who would consider these minor tradeoffs against the bigger goal of opening a new frontier and paving a safer and more comfortable path for future generations. As President Reagan once reminded the nation, frontiers aren't opened by the faint of heart.

Appropriately, the MDRS is in the middle of nowhere. I expected the Utah desert would be remote, but I was impressed that the journey to the research base required four-wheel drive, a standalone GPS, and enough daylight to navigate the dirt roads that wound sharply around jumbles of rock. Finally, just after the last flickering bar of cellular service dropped for good, I caught a glimpse of a white column emerging in the distance, dwarfed by the dramatic red canyons. That two-story, 1,200-square-foot cylindrical structure was the MDRS, my home away from Earth.

At Masten, I was working every day to mature the pinpoint-landing technologies required to safely deliver cargo, supplies, and eventually humans to future off-Earth settlements. Here at the MDRS though, I was about to experience for myself the ups and downs of life as an early settler on Mars.

My international research team included civilians and scientists from JAXA, NASA, academia, and private industry. We were educators, trauma surgeons, MEDEVAC pilots, and microbiologists, and together we immersed ourselves in a multi-week simulation of life on Mars, sharing 1,200 square feet of living space and trying our best to make a home out of a hostile environment. We arrived in Utah from vastly different backgrounds and with a diversity of research interests, but our pilgrimage to Mars was grounded in the shared belief that space settlement was an achievable goal within our lifetimes, as well as in a common desire to help accelerate progress.

That said, living and working in harmony in such close quarters required significant effort up front. In our first consensus-building exercise, we discussed as a group how we should approach the analog research stay. One option was to keep a foot in the real world, tethered to reality in small but simulation-breaking ways. The more extreme option was to completely suspend disbelief, committing to respecting the boundaries of a deadly environment and isolating ourselves from Earth. Our crew unanimously chose the latter approach. We also discussed under which circumstances we would break that simulation, and which types of medical episodes constituted true emergencies. We had a trauma surgeon on crew who could deliver field care and most of us were capable of rendering basic first aid, so it was decided that loss of limb, consciousness, or life would be the criteria for calling emergency services. We hoped that wouldn't be necessary for a number of reasons.

Our days were carefully scheduled, a rotating blur of chores and maintenance duties that left us exhausted by the time we finally climbed into our tiny rooms, which themselves offered little more than a concave sleeping slot, shelf, and desk. It wasn't much, but it was a merciful private space into which we could retreat when the strain of living quite literally on top of each other proved too much. We got along well, but also experienced the friction you might expect when seven people are asked to live and work within the confines of what would ordinarily be considered a small apartment.

Packing had also presented a challenge. What does one bring along on a journey to Mars? In addition to my research equipment, and out of respect for the simulation, I landed on a modest list of personal effects, including a journal, a fully loaded e-reader, hundreds of baby wipes, and a number of basic layers to wear underneath my flight-suit uniform in protection against the biting desert cold. I also smuggled a small bar of chocolate, a delightful contraband to be unwrapped in private and savored on my birthday, which I had committed to spending in isolation with six strangers. Time away from family and friends was perhaps the biggest challenge. Back at home, I left my confused but supportive fiancé searching for the right words to explain to our mutual friends where exactly I was disappearing to for the next few weeks.

As expected, day-to-day life on Mars was a grind. With the exception of high-tech lab equipment, the Hab afforded only the crudest of facilities. We were completely isolated from Earth with the exception of a desert rat we could hear scrabbling across the ceiling beams. On the third night, we named that honorary eighth crew member "Murphy," in honor of all the things that could and had gone wrong.

Almost immediately after settling in, we experienced a total loss of power, fuel, and communications. These catastrophic losses were par for the course on Mars, and having our toilet, refrigerator, water pump, and link to the outside world taken away kick-started our transition to self-reliance. We got to work constructing field latrines, rationing water, and using the last of our stored solar power to laboriously relay with Mission Control in an ultimately successful effort to transform a rover into a temporary power generator. And we did it all in spacesuits.

Throughout our rotation, access to the internet remained unreliable at best, and nonexistent at worst, but losing electricity was an inconvenience that paled in comparison to the emergency presented by our shrinking supply of water. Our crew implemented strict rationing of the emergency reserves for drinking and reconstituting our assortment of freeze-dried meals. Bathing was best achieved with baby wipes, and the rare luxury of a faucet shower was measured in seconds, rather than minutes. The living conditions were challenging, but they presented a reasonably good analog for what we could expect as the earliest visitors on Mars (with the exception of the rat, of course). We had to lower some of our expectations: we had arrived armed with enough research projects to sustain a small nation's space program and had underestimated how much energy and effort would be dedicated to simply staying alive. Once we had those basic daily survival protocols under control, we could finally turn our attention to the research that had earned each of us a spot on the crew.

Among my own planned research goals was a demonstration of additive manufacturing. It would be impossible to pack every single tool you could possibly conceive of needing for any given situation for the rest of your life, and I aimed to demonstrate how valuable it would be for early settlers to have the ability to produce novel tools off-Earth in a pinch. To that end, I had packed a small 3D printer, and when power was restored, the first thing I ended up printing was a mousetrap, a tool I had certainly not anticipated needing on Mars (QED).

In addition to the 3D printer, I also brought along a prototype spacesuit from Final Frontier Design, which I tested during EVAs, or space walks. In surveying the area's natural resources, we practiced the *in situ* resource utilization techniques early settlers would rely upon. Leaving the Hab is where the simulation really kicked in. The Utah desert air was obviously breathable, but we had committed to operating as though exposure to it would kill us. In respect of that danger, we ran through our EVA checklists with great care, double and triple checking our equipment, helmet seals, life-support systems, and radio communications, as well as observing a three-minute pressurization cycle in an airlock before venturing onto the Martian surface. EVAs were limited to repairs and research rather than recreation, and when we did journey out, it was in pairs of two or three, ensuring that at least two crew members were stationed back at the Hab to monitor our progress and vitals. Each venture out began with the same radio relay: *"Hab, this is EVA 149-3. Visual confirmation of six millibar pressure reading. Over." "EVA 149-3, this is Hab-Actual. Copy that, you're clear to exit. Over."*

Many aspects of the Martian environment were simulated, but the scientific research was entirely real. We used a few of our EVAs to search for lichen colonies in the desert around the Hab and collect samples. Whenever ice sheets or new land bodies form, lichens are the first settlers; they're some of the hardiest organisms on Earth. The lab afforded us use of a centrifuge, which we used to separate the lichen samples before using a sequencer to further identify extremophiles and cyanobacteria. A growing body of research supports the idea that cyanobacteria could aid in an early terraforming effort on Mars, thanks to their penchant for photosynthesis. Alone, they wouldn't have a chance in the harsh Martian environment, but resilient lichen colonies could provide a hearty housing shell that shields them from harmful UV rays.

Aside from extensive geological surveying to map out the most useful materials and collect samples of microbes, we also tested new ozone-laundering systems for dirty laundry, evaluated novel aerial photography techniques, and even conducted a dietary study focused on "insect palatability" as a clean, renewable protein source. Our crew found the relatively tasteless cricket powder to be the least offensive—perfect for protein bars. Other more exotic species, such as the boiled zebra tarantula, were predictably unappealing.

Our crew rotation happened to coincide with my twenty-sixth birthday (13.8 in Martian years), and I enjoyed the unique experience of throwing a birthday party on Mars. My crew wrote me birthday notes, relinquished me from chores, and whipped up a celebratory dinner of macaroni and reconstituted cheese. As an extra special gift, I was extended the honor of naming two plants who were part of our forced plant-growth study, and that night we christened Rosenplantz and Gildenfern in the GreenHab. As a crew-wide project, we had carried to Utah nearly fifty pounds of

Martian regolith simulant, intending to investigate the viability of sorghum seeds and hops rhizomes in NASA-grade Mars dirt.

We chose these crops for a few reasons. The official reason was that sorghum is a highly nutritious grain that has relatively low water needs, and hops can be used as a medicinal herb, making both of them suitable candidates to bring to Mars. The more exciting reason, and likely more interesting to the rest of the world: they're two of the three constituent ingredients in beer. Yeast has already been sent to space, so if we could prove germination and root establishment of sorghum and hops, we could show that one can produce beer on Mars.

As it turned out, our experimental plant group thrived in Martian soil. By the end of our crew rotation, our Hab was lush with budding foliage. This was an exciting outcome: after the plant growth, the normal brewing process could take place, resulting in an eventual keg of well-deserved space beer. With all the challenges of settling Mars, the ability to have a cold beer is a small comfort that might make early life slightly more appealing.

As it turned out, this so-called "Proof of Beer" study attracted quite a bit of media interest. In addition to filming segments for ABC's *60 Minutes* and hosting British comedian Karl Pilkington as an honorary crew member for his show *An Idiot Abroad*, a number of magazines and media organizations reached out with interview requests. I was flustered and flattered to receive a feature request from *Playboy,* only to be humbled seconds later when it was clarified that this one really *was* all about the article: the accompanying photo of me would be in my full spacesuit, visor down. I think I still hold the record for most fully clothed woman ever featured by *Playboy*.

The Martian research success was exciting, but we also exported a lot of valuable lessons back to our homes on Earth. Resource conservation is one of the most critical components of our ability to settle space, and hearing the *glug-glug-glug* of the water pump every time we turned on a faucet gave me a profound understanding of how much water we use—and waste—over the course of a normal day back on Earth.

The experience also convinced me that personality will be a huge factor in aptitude for long-duration spaceflight and eventual settlement, much more than specific technical background. Survival is a full-time job that will require extensive cross-training in a variety of disciplines, but medical and engineering skills can be learned. I emerged from my crew rotation with a new appreciation for the intangibles, like the psychological and emotional fortitude candidates will need to possess, as well as their interpersonal skills, personal values, and conflict resolution skills.

No one wants to be stuck on a deep-space mission with someone who is pessimistic, confrontational, egotistical, or otherwise disagreeable. Once upon a time, getting into low Earth orbit required a steely-eyed determinism that NASA believed could only be found in military test pilots, so those were the profiles that dominated our early astronaut classes and societal imagination. For space settlement, we're looking at an entirely different, diverse pool of candidates optimized around crew compatibility.

My crew left the MDRS convinced that human expansion in the solar system is feasible in our lifetimes, a goal that we were each dedicated to pursuing. Besides our commitment to advancing the space exploration capabilities of humankind, we also shared something else in common: if provided the opportunity, each one of us would jump at the chance to go ourselves.

That wasn't necessarily a hypothetical. At one point, every member of my crew had submitted applications to Mars One, a controversial nonprofit company proposing to send four humans on a one-way ride to settle the red planet. We all agreed that the program was extremely unlikely to materialize, but we also agreed that we, as a species, need to start making real progress toward that goal, and we were willing to hear out any organization that was trying to close the business case.

CHAPTER 7

REDEFINING THE
RIGHT STUFF

WHEN I FIRST HEARD ABOUT MARS ONE, I WAS INTRIGUED. HERE WAS A MEDIA

organization wading into a very important conversation about permanent space settlement; it was a cause worthy of further investigation. Along with 200,000 other people around the world, I sent in a video, a few personal essays, and a modest application fee. I held no illusions that the organization would be able to pull off the journey, but a global conversation and non-zero odds of moving the needle were better than nothing.

Here's the thing: Mars One never set out to be an aerospace company. They were a supernova of a nonprofit organization, exploding into the public sphere with a single media premise, attempting to close a multi-billion-dollar business case through broadcast rights. Mars One believed that the global spectacle of sending the first humans to Mars might raise enough publicity dollars to pay for the hardware to get them there in the first place. This last point resonated with my MDRS crewmates: we found common ground with Mars One in our shared belief that the major barriers to human settlement of Mars are largely economic.

The plan presented plenty of fair critiques, but I found the broader giggle factor around Mars settlement unwarranted. The coverage often ignored the actual achievability of sending humans to Mars, zeroing in instead on the perceived lunacy of anyone willing to sign up for a one-way mission. While I privately agreed that Mars One wasn't going anywhere, I considered forty dollars a reasonable investment for an opportunity to help redirect the conversation. I quickly found myself in a candidate pool down-selected from two hundred thousand to five thousand, and then further reduced to one thousand. Eventually I found myself listed among the final one hundred candidates, which put me in a precarious position. The respectable career and profile I had built for myself in the space industry was beginning to be eclipsed by the growing and controversial spectacle of Mars One.

I wavered between doubling down on principle or futilely trying to scrub any mention of them from my Wikipedia page. In the end, I decided on an approach somewhere in the middle. If people were hell-bent on giving me a platform, I was going to take it. My relationship with Mars One would deteriorate for other reasons, but before it did, I took my conviction and talking points on a brief media crusade, aiming to impart that while we should all be skeptical of bold claims from any organization, we should refocus our scrutiny on the technical aspects of the challenge itself rather than deriding the people who believed advancing humanity's footprint was a worthy and achievable mission.

Across television and print media, I attempted to explain that planetary exploration was vital for a whole host of reasons, like learning more about the mysteries of our own planet (you know, the one on which all known life exists). For those focused on the bottom line, I pointed to the proven economic benefits of flagship space programs; some estimates say that for every dollar invested in the Apollo program, more than twenty have been returned, many in the form of advancements in computer technology, communications, rocketry, satellites, surveillance, weather imaging, life-saving medical procedures, asteroid detection, and the rest of an outrageously long list of spin-off technologies that Earthlings have come to depend on in daily life.

But mostly I focused on the emotional reasons why someone might want to explore our universe: we're a species of explorers, continually seeking and settling new frontiers. The question shouldn't be why would we, but why haven't we already tried? I gave one impassioned interview in which I drew comparisons to Arthur C. Clarke, who once wrote that all new ideas pass through three stages: 1) it can't be done; 2) it can probably be done, but it's not worth doing; and 3) I knew it could be done all along.

When he first proposed a plan to utilize geostationary Earth orbit for communication satellites in the 1940s, it was an idea was firmly stuck in stage 1, because the technology seemed impossible at the time. Of course, today, with nearly our entire society dependent on satellites overhead, the idea has graduated to the obviousness of stage 3.

I shared with my interviewer an analogy that the Mars Society has long emphasized: the concept of space settlements moved past stage 1 decades ago, and ever since then, we have been stuck in stage 2, with a goal that everyone agrees is achievable but one whose incredible costs and efforts we have struggled to justify. The main message I wanted to communicate, I stressed to my patient reporter, was that stage 3 can be reachable within the lifetimes of anyone reading the article; I wasn't campaigning for Mars One in particular, I just didn't want discourse around an important direction for our species to get lost in the laughter. As we explore space, we *must* do so with the intent to stay—

space settlement is a long-term commitment toward long-term survival.

Whether in low Earth orbit, on the Moon, on Mars, or all of the above, I want to see it happen. If not for our sake, then for the sake of the next generation. I reiterated that I was fine knowing it might not be possible in the time frame that's being proposed, and once again for emphasis, that I didn't actually care about Mars One in particular. I'm perfectly fine with seeing a number of organizations try and fail as long as it moves the needle. What I would not be fine with—and now I was reaching a fever pitch—is knowing that we were refusing to take any steps to work toward that goal and dismissing any and all discussion as a joke.

A few cool-down minutes of banal Q+A later, and the interview ended. I hung up with the reporter certain that I had absolutely crushed it, delivering the nuanced and rational hot take that America needed to hear. When the article came out the following week, my picture appeared under the headline "BRIDE-TO-BE TO DUMP FIANCÉ FOR ONE-WAY TRIP TO MARS" and it didn't get better from there.

If you thought I learned my lesson after that, you would be incorrect. The same sort of fierce tenacity that propelled me into the space industry in the first place ensured that I would continue submitting myself for public humiliation, a futile attempt to refocus a national conversation. Clearly this topic was difficult for a lot of people to digest, and there were some major misconceptions, both scientific and emotional, that needed to be clarified. In print, my perspectives had been editorialized and misrepresented beyond recognition. *Maybe a video interview would be better*, reasoned the painfully naïve film major who deserved everything she had coming her way.

That's the level of optimism and audacity that found me on ABC's *The View*, staring back at a live audience while Jenny McCarthy, Whoopi Goldberg, Barbara Walters, and a Real Housewife from Somewhere peppered me with questions that derailed my talking points beyond my wildest imagination.

"So, this suicide mission..." started Jenny. "*Settlement* mission," I clarified. The rest of the five minute interview I have blacked out from memory as some sort of self-protection mechanism, but I remember being absolutely bewildered by the Real Housewife's line of questioning, which seemed to be centered around whether I had considered the risks of being stuck on board with potential sexual predators during the long journey and other contemplations that I could never have begun to prepare responses for. I stumbled through a few assurances about psychological screening and crew compatibility. After a merciful cut to commercial break I thanked Whoopi and Whoopi only, telling her I loved her work in *Star Trek*. Then I took off my mic and got the hell out of dodge.

At this point, I was finding it tedious to be on the defensive for an organization I wasn't even personally invested in, but I begrudgingly agreed to one last speaking engagement on the matter, in the form of a keynote where I could share my perspective uninterrupted to an audience of peers who I believed could fairly evaluate the nuance of my position. The invitation came directly from the Federal Aviation Administration, and it was at their annual Commercial Space Transportation Conference where I took a deep breath and delivered my very last missive on Mars One.

FAA Commercial Space Transportation Conference
Spoken Remarks—Kellie Gerardi

Thank you to the FAA for inviting me here to speak today. I've developed a productive relationship with the Office of Commercial Space Transportation over the last few years through my work with the Commercial Spaceflight Federation, the DARPA XS-1 workshops I ran through the Space Frontier Foundation, and my more recent move to Masten Space Systems. When asked to speak, I wasn't quite sure what topic you had in mind for me, but I was confident I had plenty of industry material to draw from.

So when I looked at the draft schedule and saw "Kellie Gerardi—Mars One," I ran through a couple of emotions. First, I felt a little bit embarrassed, like a flash-in-the-pan YouTube star presenting an award at the Oscars. I also felt some professional concern—is this the image I've cultivated for myself? Not as a respected industry professional but as a tinfoil hatter planning to hitch a ride to the red planet? Those reactions were based on my recent media exploits on the same topic. But truth be told, I'm glad to have the opportunity to speak about a topic that's been a punchline in conference hallways, making murky the boundaries between space people and non-space people. I'm pleased that the FAA's Office of Space Commercial Transportation, like me, saw that there's something else going on here, something that deserves some constructive discussion. I want to explore what movements like Mars One mean for the commercial spaceflight industry.

I remember when the applications first opened. An international organization who became an overnight sensation in claiming it was going to put four human settlers on Mars by 2024. Anyone could apply for a token application fee of about forty dollars. It reminded me a little bit of Ernest Shackleton's earliest expedition advertisements. It was very easy to read the headline and disregard the premise as disposable gimmick. Funding it through a reality show? Yeah, okay. Then a couple hundred applications popped up on their website. The applicants were earnest in their public-facing videos. They were students, parents, army reserve officers. They wanted to participate in humanity's next giant leap. Then those applications grew from a few hundred to a few thousand, and then a few tens of thousands. One week before the application deadline, there were over 200,000 applicants—that's almost a quarter of a million people who were interested in a human colony on Mars. Mainstream media had descended on the story at this point and it was my fiancé who was most surprised to hear that I hadn't applied. He said "This is basically your dream, right? Hundreds of thousands of people talking about settling Mars. I would have thought for sure you'd be all over this." I applied that night, because he was right. My own hypocrisy astounded me. I was already the bane of my friends' existence at dinner parties, ruthlessly monopolizing conversation with news of the commercial space industry, practically begging people to care about space. Here's a subject I care so passionately about, whose mission I have made a career out of communicating, and I'm going to sit on the sidelines while a nonprofit organization activates a quarter of a million people as impassioned spaceflight advocates and stakeholders? No way I was going to let my skepticism prevent me from having a voice in this conversation. My own application crept in right before the deadline.

Me telling you that space exploration is vital to our nation would be preaching to the choir, so I'm going to skip that. That's why we're all here today, and every day—because we share that fundamental mindset. And on some level, I think nearly all of us would jump at the chance to go ourselves. That's what I love about the commercial space industry, and why I chose a career here. This is the industry that will give me a shot at going to space myself. This is also the industry that's going to expand Earth's economic sphere. Companies here today are lowering the cost to orbit and democratizing access to space. That's the message I continue to spread in my role as media specialist of the Commercial Spaceflight

Federation—this isn't just about space; this is also about long-term high-tech human progress. This is about traction and job creation in new markets, ones that are real and growing.

I'm optimistic about the future of our species, but I also recognize that life on Earth has an expiration date. Without space settlement and advanced life-support systems, our entire species does as well. On a more emotional level, I want humans on Mars simply because it's within our reach; the entire solar system is within our reach. That's the mantra that makes me proud to go to work at Masten every morning, knowing that we're maturing the technologies that will unlock access to the unexplored corners of the universe.

At some point, whether in the near future or very far, we know that either planet Earth will cease to support life or humanity will render the planet uninhabitable. This is the first time in over four billion years that it's possible for life on Earth to become interplanetary. Otherwise, as Professor Stephen Hawking puts it, "we're like castaways on a desert island." So when someone talks about putting humans on Mars, or on the Moon, or on literally any other surface beyond the International Space Station, I am desperate to see it happen.

I want to pull the curtain back on Mars One a little bit. They have a nonprofit organization and a single media premise. They're trying to do this in the most nontraditional way possible, and they'll be the first to tell you that they're not an aerospace company—they're simply trying to close the business case. And you know what? They're doing something significant. In the same way that finely crafting a subsystem component is an incremental step in achieving spaceflight, so too is this global Mars One conversation moving the needle on space settlement. That's the same spirit motivating people like me to contribute and invest my time in this conversation—I have seen firsthand the ability to activate people's imaginations.

As the media specialist of the Commercial Spaceflight Federation, it has been my job to support the democratization of space, the expansion of Earth's economic sphere, and humanity's ever-reaching journey in space. I'm accustomed to our industry reaching this level of media attention only when something's gone horribly wrong, so I'm grateful to Mars One for providing me with a platform to speak out about all the things that are going so well in our industry. Over the past year and a half, I've had the opportunity to appear on ABC's *The View*, *Nightline*, NPR, and a VICE Motherboard documentary. I've contributed to articles in the *New York Times*, *Popular Science*, *Vogue*, and more. The conversations haven't all gone the way I would have liked, but I've had the opportunity to inspire imaginations and spread positive news about the commercial space industry. It's been refreshing to reach millions and millions of people—some of the same people who thought space exploration ended with the shuttle program—and share our remarkable progress with them.

I can attest that the majority of the general public doesn't look at space travel, technology, or exploration as anything other than scientific curiosity. And why would they? There's an almost unconquerable divide between the way we see space and the way they do. And that's because there has not been a meaningful translation of what space could mean for them. The argument of potential spin-off technologies isn't enough. Unless they see opportunity in space, the potential to positively affect their lives, we don't get that level of public engagement and support.

I look at Mars One applicants and I see thousands of people who have never taken an astronomy or engineering class in their life suddenly obsessed with space exploration. I see people around the world studying up on *in situ* resource utilization and planetary protection concerns. Better yet, I'm watching waves of people beginning to take an active interest in the space activities actually happening around them. *Your* activities. They're online celebrating SpaceX launches, they're retweeting images from the ISS, they're championing Virgin Galactic, and they're spreading that infectious excitement for space exploration far outside of our industry echo chamber, and into their own communities that were previously inaccessible to all of us. We've welcomed hundreds of thousands of people, and all of their personal friends and families, as activated stakeholders in in our industry.

Putting aside the feasibility of their launch and mission plans, Mars One has already succeeded in accomplishing something extraordinary. They've managed to convince almost a quarter of a million people to look at space through the lens of what it could mean for their personal lives. That awakening doesn't end with Mars One.

What at first glance looks like a laughable selection process suddenly seems more profound, the very embodiment of democratizing space. We're broadening the definition of what it could mean to be an astronaut and we're redefining how a generation looks at space. If we're going to sell the vision that all you need to be an astronaut is one or two hundred thousand dollars, then we damn well better be prepared to explain to one or two hundred thousand people why that doesn't apply to their dreams, why that logic isn't scalable. It's in our best interest to look at these aspiring Martians and not scoff at them but encourage them. It's our job to confirm that vision of what a collaborative future in space could look like. And then it's our job to engineer it into reality.

I think Mars One applicants are largely misjudged. This isn't some group of planetary malcontents waiting around for their rocket to Mars. This is largely a group of educated professionals whose vision for the future isn't crippled by the barriers of the present. Much to my surprise, I've discovered an overwhelming number of early applicants were also space industry professionals, some of whom are in this room today. The overwhelming motivation is a shared goal of expanding human presence in the universe. They don't care if it's Mars One, and they don't care if it's them personally. Space settlement is a goal they want to see developed for humanity, period. And they're doing everything they can, professionally and personally, to further that goal.

I'm too young to give up on that kind of future. I'm too enthusiastic to sit around and let other people handle the difficult and extraordinary things. I'm too committed to suggest that someone else should solve the problem that I'm not willing to contribute to myself.

I'm also old enough to not believe my own hype. I have a full-time job at Masten Space Systems. I have bills to pay. I know the future won't be quite as awesome as my unrestrained imagination and optimism, but I'm committed to making an honest, rational effort. Mars One was never an aerospace company. They were an organization trying to creatively close the business case for humans on Mars. And for that, I applauded them. Their goal was simply to raise the money to pay industry to design, build, and fly the hardware.

I've found that generally, people accept the core feasibility, and public interest lies more largely in the philosophical considerations. Would you really want to leave your friends and family forever? You would be okay with dying on Mars? Those softball questions bring me indescribable joy. I'm thankful to be alive in a time where the stigma of an emotional barrier has surpassed the engineering barrier.

At Masten Space Systems we have a team of masterful engineers who know you don't have to be two hundred miles above Earth's surface to change the future of space exploration. For us, you just have to be two hundred miles north of Los Angeles, in Mojave. I'm proud every single day to know that the pinpoint-landing technologies of our rockets will help to unlock the unexplored corners of the universe. If you ask any of our engineers why they work at Masten, anyone on our team will be eager to spell out the implications of our VTVL vehicles flying Mars-like descent trajectories and autonomously detecting surface hazards and diverting to a safe landing location. They'll explain how the pinpoint-landing capabilities of our rockets are the difference between miles and feet, the difference between landing in a desert near a Martian crater and landing safely on the crater rim itself.

Detractors might point out that our operation in Mojave is as far removed from space settlements as an initiative like Mars One, but that's shortsighted. We do detailed, technological work, and we're making incremental progress toward unlocking access to the most scientifically interesting corners of the solar system. We all share NASA's strategic goal of expanding human presence in the solar system, but as revenue-generating companies, we also know our limits. As badly as some of our engineers might like to prove out their other ambitions, we must first find customers for those services.

So when an organization comes around and says they're going to do their best to close the business case, to raise enough money to pay companies like Masten, SpaceX, Lockheed, Paragon, and all of the rest of us to do what we do best, I give them my full attention. As an industry we need a broad range of approaches and visions. Destinations like Mars need both pioneers and patrons. We have a few, but we're going to need a lot more. If organizations like Mars One can give the public a dream to latch onto, it might just help turn a conversation into a convoy.

If I were defending Mars One, I would be asking you to consider the history of Mt. Everest, from its official naming in 1875, to the first attempts to climb it almost fifty years later. I would argue that less was known about the summit of Mt. Everest in the 1920s than is known about the surface of Mars today. Want to guess how George Mallory was able to fund such an expedition? Through the sale of video rights and the creation of one of the world's first documentaries.

If my goal were to defend Mars One, I would give you ten more historic anecdotes of related precedence. But it's not my job to defend them or any other specific organization. It's my job to propel the capabilities of the commercial spaceflight industry. If some organization— any organization—closes the business case, and there's enough money to contract the commercial industry to begin phased work and feasibility studies, it's my job to defend the fact that we can deliver. It's in our best interest to defend that space colonization faces greater economic barriers than it does engineering barriers, and I want the public to understand and appreciate that difference.

It's dawned on me that we need more than one tool in the toolbox to construct this future. If we sit around waiting for the 1968 mission plan to follow up Apollo, it's not going to happen, or at least not for any of the people in this room. It's been almost fifty years. I'm prepared to fight for the chance to see this in *my* lifetime. And that means the budgets can't be tied to an election cycle. This is a long-term commitment to our future, and we're going to need external ideas, financial commitment, and international public support to make that dream viable.

When something like Mars One crops up, it's easy to throw away the premise. It's easy to say, "that's ridiculous" and dismiss any further consideration of the topic. Some of us are reacting like this is the first time anyone has proposed an unproven idea to the space industry. But this is something that deserves attention. New attention. Big outlandish ideas are largely destined to fail, but even a moment's consideration can have a profound impact on progress toward shared goals. Mars One has already succeeded in gaining the attention of a broader audience. Not only this audience of my peers, but a broader audience of people who once thought NASA was canceled. Human occupation of Mars is having a significant moment in the public spotlight. Mars One is unlikely to ever advance beyond a conversation, but that conversation alone is narrowing the gap between science fiction and science and elevating the discussion to a global scale.

We as an industry are presented with a chance to direct and define these kinds of movements. We can sit on the sidelines with our arms folded, irrationally irritated that thousands of "non-space people" are talking so unreservedly about space settlement. True, these are people who haven't sat through years of panels on export control or launch indemnification, people who are most likely blissfully unaware of ITAR restrictions or learning periods or experimental permitting. They may not even be aware of all the interesting and terrifying ways they could die on Mars. We can dismiss it as superficial advocacy of space exploration and deride the applicants.

Our other option is to lean in as guiding force and voice of authority. To say, "Yes—this can be done, this needs to be done, and here's *how* it should be done." Whenever

I hear someone say, "Mars One is a scam," it doesn't sit right with me. It's not that I care about the reputation of Mars One, but rather that I care about the implications to the commercial space industry. That person who calls Mars One a scam has failed to understand that to call a proposed Mars settlement mission a "scam" is insinuating that we as an industry aren't capable of executing this. And I happen to think that we are.

I'm excited by the achievable goal of space settlement. I'm excited that my comparatively short lifespan on Earth just so happens to fall within a Space Age—the first window in history where interplanetary life is possible. As of today, right now, I'm done speaking about Mars One, but I'll continue speak out loud my desire to be part of humanity's next giant leap, however it happens. This is the first time we have the opportunity to settle other worlds, and I want to do everything in my power, both personally and professionally, to make sure we don't waste it. My career with Masten Space Systems advances the pinpoint accuracy necessary to land rockets safely at off-Earth settlements. My personal research at the MDRS gave me a firsthand idea of what sort of *in situ* resource utilization techniques those settlements might entail. And my participation in Mars One gave me a platform to communicate all of those things to the world.

Thanks.

CHAPTER 8

FOR ALL OF HUMANITY

I MENTIONED THAT I CAME TO HAVE A COMPLICATED RELATIONSHIP WITH MARS ONE, but perhaps fractured is a better description. I continued to defend the vision of space settlement on principle, but I soon quietly distanced myself from the organization (and that was long before their cash and media captivation fizzled out).

My public speaking crusades had attracted the attention of the Dutch entrepreneur and cofounder of Mars One. He saw in me a credible advocate for the feasibility of space settlement, and we occasionally traded emails on various press opportunities or potential technical partners. When he visited New York City, I offered him a tour of the Explorers Club's venerable headquarters, and we finished with lunch nearby. We discussed a wide range of topics, musing about potential crew compatibility and health considerations and somehow, over plates of overpriced cheeseburgers, we arrived at the topic of career growth and corporate hiring. He casually shared his suspicion that while American leaders may not admit it, most hiring managers would never hire a twenty-six-year-old woman when presented with the option of an equally qualified twenty-six-year-old man. He took another bite as I, a twenty-six-year-old woman, stared back at him. He mulled over the unfortunate but inherent investment risk, in which the woman presumably gets pregnant and leaves the poor hiring manager right back where he started.

If this was an opening to discuss my family planning intentions, I wasn't biting. The redness blotching my cheeks wasn't because my feelings were hurt (please trust I have heard worse) but because it was dawning on me that this was the sort of sexist crap that could potentially be exported to another planet, or worse, prevent large swaths of humanity from getting there in the first place. Also, for all the rhetoric about kick-starting a new generation in space, any discounting of the child-bearing half of the species seemed like an enormous blind spot.

I don't believe he intended any offense; rather, it was a simple business logic that he suspected drove managers to eschew female candidates inside of the most common reproductive red-zone years. But it echoed the frustratingly defeatist sentiment that *Mercury 7* astronauts had deployed decades ago in testimony against the inclusion of women in spaceflight, that "desirable or not, it is simply a fact of our social order" and that "we are only looking to people with certain qualifications. If anybody can meet them, I am all for them..." Above all, it was a powerful reminder that we need to be intentional about the values we carry with us into future worlds.

After that conversation, I inspected my industry with a sharper lens, hyper alert to the importance of diversity, inclusion, and equity in the context of space exploration. I was accustomed to being the only woman at the table, in the meeting, on the panel, or in one case, at the entire company. Negative experiences that I had once assumed were just the reality of working in aerospace and defense now struck me as critical problems that risked our industry's overall success. Inclusive language was perhaps the lowest-hanging fruit. Along with NASA standards, I had long swapped out "manned flight" for "crewed flight" or "human spaceflight" in all of my speaking and writing. Next, I made it my mission to ensure that outdated language didn't persist on

Wikipedia pages and in headlines. I also pledged not to moderate or introduce any more all-male panels (manels?), something I had recently begun to recognize as a contributing problem and demotivating pattern in my conference circuits. If they couldn't find any female experts to include on the panel, then it was an invitation I politely declined.

But if the industry gender breakdowns were bad, the racial disparities were downright terrible. To paraphrase Douglas Adams, space is white. You won't believe how vastly, hugely, mind-bogglingly white it is. This presents not only an access problem, but also an experience problem. Bias, both overt and implicit, is a societal plague from which the space industry is most certainly not immune.

The impacts of racial and gender inequities are profound and far-reaching, especially in the context of space settlement. When we talk about space exploration and space settlement, we're really discussing the future of the human species—and it's critical that that conversation includes a diversity of perspectives and vested stakeholders. If the space industry wants the world's trust and support in this quest to open the final frontier for the benefit of all of humanity, then the space industry needs to more fully represent humanity. There are a number of giant leaps that need to happen to ensure we're engineering an inclusive future, but it all starts with making sure a diversity of voices is represented now, as we begin to set course, and holding our peers and leaders accountable for prioritizing it.

Of course, there are resisters, even among my peers. There are some for whom "diversity and inclusion" is a triggering phrase in itself, and plenty who would scoff at the notion, clinging to a flawed meritocracy, and insisting, "we only want the best." All despite the fact that we're an industry full of advocates who pride ourselves on democratizing access to space and lowering not only the cost to orbit but also the barriers. We, more than any other industry, should understand that lowering the barriers isn't the same thing as lowering the bar. Now more than ever, the space industry must realize that it can never exist independent of what's happening on Earth—space has always been inseparable from society. Humanity's brightest future in space starts with investment in people right here on Earth; looking only upward while refusing to look inward would be nothing short of betrayal.

It's also shortsighted. As spaceflight evolves, so too will crew needs and ideal candidate profiles. Mars is a perfect example— we've never sent someone so far away from their home planet and the rest of their species, and we've spent very little time studying the vastly different set of psychological challenges that those candidates will be required to confront. And if those earlier pioneers travel with the intent to stay and settle a new world, we need to think carefully about the value system we want our earliest astronauts to export to that world. We need to rethink their qualifications through the lens of compatibility, partnership, personal values, and emotional fortitude. If the steely-eyed test pilot was considered the ideal candidate for a high-octane orbit of the Earth, who is considered the ideal candidate to make a home out of Mars?

And we'll soon need to think even more broadly. If we are to survive as a species in the long-term, small cohorts of hand-selected, highly trained astronauts aren't going to cut it, no matter how diverse their crews are. We need to approach the problem with a more universal goal in mind: democratizing access to space for humans of *all* backgrounds and abilities, even—no, *especially*—regular people like you and me. (And if the species is to continue reproduction in its current form, then yes, even women of child-bearing age.)

To trot out a tired phrase, it's not a nicety—it's a necessity. Diversity in any workplace means diversity in thought and novel approaches to complex problems. In high-tech, high-stakes industries like aerospace and defense, it's in everyone's best interest to cultivate the broadest possible set of perspectives.

After many years of whiteboards and war rooms, I had learned that the people designing technology hold the power to influence how it's applied. With our species on an astronomical trajectory, the stakes are too high and the impacts too far-reaching for any single demographic—or country—to steer the entirety of Spaceship Earth, however well-intentioned they may be. In fact, unlocking the future has always been a delicate exercise in consensus-building.

Perhaps surprising given the Space Age's Cold War-era beginnings, international cooperation has long been a priority. Ten years after the Soviet Union launched Sputnik and two years before the United States landed on the Moon, global leaders sat down to hammer out some early and important details regarding the future of space travel. The Outer Space Treaty of 1967 established not only a foundational legal framework for space activities, but also a set of philosophical principles for all of Earth's spacefaring nations:

PARTIES TO THIS TREATY,

INSPIRED by the great prospects opening up before mankind as a result of man's entry into outer space,

RECOGNIZING the common interest of all mankind in the progress of the exploration and use of outer space for peaceful purposes,

BELIEVING that the exploration and use of outer space should be carried on for the benefit of all peoples irrespective of the degree of their economic or scientific development,

DESIRING to contribute to broad international cooperation in the scientific as well as the legal aspects of the exploration and use of outer space for peaceful purposes,

BELIEVING that such cooperation will contribute to the development of mutual understanding and to the strengthening of friendly relations between States and peoples,

[...]

HAVE AGREED ON THE FOLLOWING...

The treaty goes on to spell out some very reasonable guidelines, including a ban on weapons of mass destruction in space, the agreement that no one nation can claim ownership of any moon or celestial body, and a commitment to facilitate and encourage international cooperation in all the amazing space efforts to come.

The spirit of international cooperation is strong, but it often proves difficult to implement in practice. Many American companies are at the vanguard of the final frontier, but their talent pool has a major limiter: US rocket companies can only employ US citizens and, in some cases, permanent residents. The restriction is due to International Traffic in Arms Regulations (ITAR), a regulatory regime requiring that access to any US military technology be limited only to US persons and companies. What on face value seems like a reasonable national security mandate soon becomes a mess when unpacking exactly how broad and far sweeping the term "military technology" is.

Rocket technology has come a long way from the V2. Beyond the fact that they both go up and release payloads, commercial space vehicles and weapons of war have clear distinctions in both design and intent. Unfortunately, the United States Munitions List hasn't evolved to meaningfully distinguish between the two. In the mind of US law, even peaceful spaceflight technologies are viewed as potential ingredients of a weapons program, and thus restricted by ITAR. In practice, this means that cutting-edge space companies like SpaceX, Blue Origin, Virgin Galactic, Virgin Orbit, and so many others have their hands tied when it comes to hiring outside of the US.

These outdated restrictions are perhaps the biggest regulatory obstacle to the democratization of space and expansion of Earth's economic sphere and are ironic when considering the personal backgrounds of the space industry titans leading the charge. The adoptive father of Blue Origin CEO Jeff Bezos emigrated to the United States from Cuba as a teenager; Elon Musk was born and raised in South Africa, educated in Canada, and received his US citizenship the very same year he founded SpaceX; and Virgin Group Founder and space tourism visionary Sir Richard Branson remains a British citizen. This group alone shows the limitations of looking only to the United States when it comes to big dreams and bold innovation, each one leading history-making companies that would surely benefit from the opportunity to attract candidates from a broader talent pipeline.

And the United States is far from the only player in space. In addition to American astronauts, Russian cosmonauts, and Chinese taikonauts, a number of other nations have leveraged the space infrastructure of those more established space programs to send astronaut envoys of their own. Japan, Israel, Canada, Saudi Arabia, and a number of member states of the European Space Agency are among the nations who have sent representatives to orbit, earning entry to the elite club of spacefaring nations. In addition to their own national space agencies, many of these nations invest in their own academic and commercial sectors as well.

Still, space is disproportionately prioritized across the globe, a source of frustration for aspiring astronauts or aerospace engineers who don't have access to the same employment opportunities as Americans. I hear often from students whose countries don't yet have a national space program or in-space presence, and I've always encouraged them to look at it as an enormous opportunity to build advocacy from the ground floor. Whether a small satellite or an astronaut representative, there exists the opportunity to help drive an entire country toward a history-making national "first." From joining or creating a chapter of an international advocacy organization, to attending the myriad of international educational opportunities that exist across all age groups (from Space Camp to International Space University), or even kick-starting a high school robotics or cube-sat club, anyone around the world has the potential to make a difference in space exploration.

I count a number of international space leaders among the role models I find most inspiring. As a frequent speaker at global conferences, I've had the honor to meet impressive folks from all around the world, but it was an encounter with the all-girl Afghan robotics team champions that stuck with me most profoundly. The talented group of teens had overcome a number of obstacles on the path to victory, including a 2017 fiasco where they were denied visas to enter the US for a robotics competition, until public outcry forced an intervention that saw their visas approved just days before the event.

It was incredible to bear witness to the determination of this team in spite of adversity. These girls grew up in a war zone—one where educating girls was prohibited as recently as 2001, when the country was under Taliban rule—and yet they committed themselves to world-class STEM advancement amidst a backdrop where many of their families and friends continued to face violence.

Fatemah Qaderyan, the team captain, wrote a poignant essay in which she credited her father as her biggest supporter who helped her show the world that, if given the chance, Afghan girls could do anything. Tragically, just days after returning home to Afghanistan with her medal, Fatemah's father was murdered in an ISIS attack at his mosque. Despite backlash from some in her community, Fatemah remained on the team and continued to lead them, and her country, to the stars. "These girls know something that many adults struggle to grasp," their team chaperone told me. "Space is for everyone."

While access to a thriving space industry certainly makes it easier to get involved in humanity's next giant leap, stories like these prove that space engagement is borderless. Plenty of advocacy organizations boast major international participation and citizen volunteer opportunities, including the Planetary Society, the Mars Society, the Space Frontier Foundation. These organizations elevate space exploration to a global conversation and help bridge the gap between the scientific community and the general public. Best of all, participation in these initiatives can often be entirely virtual.

There are no barriers to space activism. Being a passionate and informed advocate for space exploration doesn't require any specific degree or career; it simply requires a passion for advancing humanity's progress in the universe. Some of the most influential members of space advocacy organizations have full-time careers in areas outside the space industry and choose to channel their diverse talents and expertise to the common goal of igniting passion for space. Of course, once involved, plenty of people are eager to make space their full-time job.

Luckily, the idea that impact in the space industry is limited solely to engineers and PhDs is just as flawed as the assumption that space exploration is an American-only pursuit.

While "astronaut" is the space industry's poster-child profession, it's only the tip of the iceberg when it comes to careers. Just as the Renaissance implies a cultural movement, so too does the Space Age, and we're just now beginning to see how that new way of thinking manifests across the diverse areas of art, science, literature, medicine, law, and so many others.

The International Space Station is a feat of engineering, but its ability to sustain and celebrate continuous human occupancy can be credited to a broad range of talents on the ground, some of which might surprise you. Before an astronaut launches to the ISS, chefs and nutritionists have created (and taste tested!) menus to optimize their dietary and vitamin needs, physiologists have designed their daily exercise regiments, linguists have tutored them in Russian language skills, scuba diving instructors have trained them underwater for future spacewalks, graphic designers have created the crew's mission patch and press kits, photographers have captured their portraits, producers have begun broadcasting the action on NASA TV, and public affairs and social media specialists have further amplified the message to the broader public.

These folks exist as part of a broader, enabling NASA workforce of project coordinators, administrative assistants, security teams, technical writers, quality control specialists, web designers, financial analysts, legal experts, IT teams, HR professionals, government liaisons, historians, and so many more.

In the broader commercial industry, nontechnical space careers have developed in the diverse areas of government affairs, law, sales, computer science, marketing, design, communications, recruiting, and events planning, to name a few. SpaceX famously hires baristas to launch and grow coffee bars inside their corporate facilities; Virgin Galactic employs both astronaut trainers and experienced managers; Blue Origin has dedicated teams for their educational STEM outreach initiatives; Zero Gravity Corporation has employed microgravity flight attendants; and Space Adventures and Axiom staff serve as space travel agents to help broker privately funded flights into orbit.

No matter your skill set, there's an area for you to contribute to space exploration. We need architects to design new habitats and space stations, and interior designers who can turn them into a home away from Earth. We need designers to bring to life the next generation of versatile spacesuits, and medical professionals who can weigh in on advanced life-support systems. The final frontier needs artists and authors who can inspire, journalists and photographers who can report and capture, and educators who can mold the next generation of students into adults who are motivated and capable of contributing to their future.

If the current diversity of space careers comes as a surprise to you, just wait until the in-space economy takes off. In a world of asteroid mining, in-space manufacturing, and on-orbit tourism, we can imagine the next generation of construction, customer service, and hospitality: astronaut-prospectors managing mining operations, cruise-ship-turned-spaceship entertainers, and so many more professions ripe for space-ification.

Of course, advancing humanity's ability to engineer that future is a spectrum, not a binary. There are plenty of folks fulfilled by their current professions or tied up with other responsibilities who are still motivated to contribute in small but meaningful ways. At the center of this passion is an ability to picture yourself in the future, and more often, picture yourself in space. For the growing diversity of space careers here on Earth, the allure of the astronaut remains inescapable. For me, working to democratize access to space means simultaneously laying the groundwork to get there myself.

CHAPTER 9

ONE-HANDED
PUSH-UPS AND OTHER
FEATS OF STRENGTH IN
MICROGRAVITY

living and working in space, but one of my biggest takeaways was that everyday citizens can have an enormous impact on space science and technology. In a world where ideas, content, and even business capital can be crowdsourced, why not apply that same model to science? As it turns out, NASA has long been hip to the idea, and the agency maintains an entire website devoted to citizen-science projects where anyone with access to a cellphone or a laptop can collaborate in scientific research.

From detecting asteroids and hunting exoplanets to processing and color correcting images of other worlds, eagle-eyed enthusiasts have always been invited to comb through troves of NASA data to identify patterns and anomalies. Collaboration with an enthusiastic public enables NASA scientists to prioritize their time and attention on the most scientifically interesting images or geological features among the overwhelming volume of data returned from telescopes, probes, and satellites. For citizen scientists, the engagement offers an active role in space exploration and the opportunity for genuine discovery. In many cases, citizen scientists are the first to view an angle of Mars or Jupiter never before seen by humans, or perhaps even more exciting, identify a brand-new planet beyond our solar system.

Interacting with images returned from JunoCam, the camera system aboard the Juno spacecraft, was my earliest foray into citizen science with NASA. The mission provided not only the first ever close-up shots of Jupiter's poles, but also a number of upvoted geological features. Through the online JunoCam community, amateur astronomers were invited to upload their own telescopic photos of Jupiter to help NASA decide what points of interest the powerful JunoCam should image on the gas giant. Once images of those top-requested features were captured, NASA shared the raw files online and allowed users to complete their own image processing and color correcting before uploading the finished results to a community message board. I didn't yet have a telescope, but NASA's website pointed me to free software where I quickly learned to process raw JunoCam images into Jovian works of art.

Of course, for all the digital means of contributing to scientific research, there remained plenty of offline opportunities as well. Physical research wasn't going anywhere; if anything, it was expanding more than ever before. Increasing access to space and lowering launch costs translated to new and exciting opportunities for students, scientists, and researchers.

Having immersed myself in the space industry, there was no escaping the dream of experiencing spaceflight myself. Conducting research at terrestrial analog facilities had been an excellent substitute, but I craved the real sensations of space travel. Luckily, we were fast approaching a watershed moment for human spaceflight. The emergence of reliable suborbital space tourism meant those same platforms would soon be accessible for science as well. Through grants from NASA's Flight Opportunities program, classrooms and university teams could send miniaturized satellites (CubeSats) and other experiments to be tested on commercial vehicles. And once those vehicles started carrying people, research teams would be able to fly alongside their experiments. To serve as a payload specialist was the ultimate citizen-science campaign and an ambition that had long taken root in my imagination. To my delight, I soon discovered a talented group of researchers assembling a crew for exactly such a mission.

Like most research projects, Project PoSSUM (Polar Suborbital Science in the Upper Mesosphere) was born out of scientific curiosity. In this case, the curiosity centered around noctilucent clouds, the highest clouds in Earth's atmosphere. Forming nearly fifty miles above Earth's surface in the polar summertime, the mysterious clouds are believed to be sensitive indicators of atmospheric change. In recent decades, noctilucent clouds have appeared brighter, more often, and at lower latitudes than ever before. Because the icy clouds require cold temperatures and the presence of water vapor to form—properties tied to carbon dioxide and methane—scientists hypothesize that their presence may be directly tied to human-made causes of climate change. Studying noctilucent clouds in our own planet's upper atmosphere also has the added benefit of helping us better model and understand high-altitude, low-density clouds on other planets, like Mars. I was sold.

Studying aeronomy, the physics and chemistry of the upper atmosphere, wasn't quite as foreign to me as studying Zulu, but it was pretty close in terms of a learning curve. I was surprised to discover how little research existed around Earth's mesosphere, the atmospheric zone directly above the stratosphere and right below the thermosphere. The last (only?) time I had seen or heard mention of this middling layer had been as part of a one-off middle school memorization exercise, but there were plenty of earthly mysteries to be solved at this often-overlooked altitude. Realizing how little some of the smartest scientists in the world knew about the mesosphere made me feel a bit better. More importantly, it showed me the enormous opportunities that exist to make direct and meaningful impacts in planetary science right here on Earth.

I learned that part of the reason the mesosphere is so poorly understood is because we haven't had a great means of reaching that middle zone. Balloons and jet planes aren't able to fly high enough to reach it and orbiting spacecraft can't dip low enough for close study. Suborbital spaceflight was the Goldilocks solution that held the key to unlocking this unexplored layer of Earth's atmosphere, allowing researchers to fly directly through it and capture images, measurements, and samples during both launch and reentry. In partnership with the International Institute for Astronautical Sciences (IIAS), Project PoSSUM's Scientist-Astronaut Qualification program was born to bring this science mission to life.

Under the leadership of Dr. Jason Reimuller, a physicist and former NASA systems engineer, IIAS grew quickly into a world-class aeronautics research and education program, training scientists for the suborbital research flights that would enable a comprehensive study of our atmosphere and the role it plays in understanding our global climate. From the beginning, Dr. Reimuller focused on attracting scientists and STEM professionals from all over the world; he knew that citizen science could engage and leverage the international science community to compound NASA research.

Qualified candidates would be trained to fly and operate equipment as part of NASA-supported research, like the Polar Mesospheric Cloud Imaging and Tomography Experiment. As interest grew and capabilities were validated, IIAS prepared their scientist-astronaut candidates for an even broader range of space science, including bioastronautics. To further the body of research around microgravity and support the development of novel life-support systems, they partnered with Final Frontier Design, whose next generation spacesuits were supported through a NASA Space Act Agreement. Scientist-astronaut candidates would evaluate and test those spacesuits in microgravity environments, high-G environments, analog flight, landing, and post-landing environments.

The barriers to space were crumbling to make way for a new generation of researchers, commercial astronauts, and payload specialists. This is what it meant to live and thrive in a Space Age, and I wanted to get in at the ground floor.

One of eight candidates selected for the 2017 Scientist-Astronaut Qualification program, I packed my bags for Daytona Beach, Florida. I had given birth to my daughter less than four months before, but I had recovered quickly, and an FAA third-class medical certificate had cleared me for training. The months prior had been a blur of baby bottles, diaper changes, and webinar-based instruction on climate science, remote sensing, and celestial mechanics. Now it was time for the in-person training, an intensive program focused on high-altitude and hypoxia awareness, high-G and microgravity endurance maneuvers, spaceflight simulators in fully pressurized spacesuits, aerospace physiology, and instrumentation operations. Once successfully completed, I would be a qualified candidate to fly airborne and suborbital spaceflight research missions.

I was beyond excited to get back in a spacesuit, no less one from Final Frontier Design, whose prototype EVA suit I had tested at the MDRS. Now I would have the opportunity to test the next generation intra-vehicular activity (IVA) suit, developed specifically for use during suborbital spaceflight. From a classroom at Embry Riddle Aeronautical University, my class of eight would pick up where the webinar course had left off, diving even more deeply into aeronomy, noctilucent cloud science, solar mechanics, and crew resource management. In addition to classroom instruction, hands-on training would equip us for the physical and operational rigors of suborbital space research.

High-altitude training was first. In the classroom, our crew learned about the perils of cabin depressurization. An explosive decompression would be fatal, and as the term suggests, would occur too quickly for air to safely escape from your lungs. A gradual decompression is survivable with proper course correcting maneuvers, but only if you recognized it happening in the first place. The issue with gradual decompression is the potential for it to occur so slowly that you may not even notice it before hypoxia sets in. The best way to prepare for such an emergency would be to experience slow onset hypoxia ourselves, so that we might recognize our own individual warning symptoms.

To induce hypoxic conditions, my crew headed to the Southern AeroMedical Institute, where I spent the morning piloting an aircraft simulator from a hypobaric chamber. Hooked up to equipment that monitored my oxygen levels, I steadied my joystick as my oxygen supply dwindled. Climbing in at sea level, I was bright and alert, but by the time we reached the simulated equivalent of 25,000 feet—just below commercial aircraft altitudes—my oxygen had dropped below 70 percent and my communication with Air Traffic Control had become noticeably delayed. My tongue felt heavy and forming words exhausted me. I was still in control of my aircraft, but I felt a flush of facial warmth and fatigue, which I dutifully reported. "Excellent, time to come back down," crackled a voice from my headset. My oxygen rose back to normal as I descended, and the symptoms melted away.

Over the following days, our high-G training took us to the actual skies, flying under the supervision of legendary aerobatic pilot Patty Wagstaff and her team. Her fleet of specialized two-seater aircraft were capable of sustaining significant aerobatic stresses, and through barrel rolls, pull-ups, and nosedives, we learned to endure the three main types of G-forces that concern aviators and astronauts.

First up was positive Gs, an "eyeballs in" acceleration force that forces blood away from the head and toward the feet as the airplane or rocket pulls up against gravity. During a space shuttle launch, astronauts would typically experience a maximum g-force of around 3 Gs, or the equivalent of three times the force of gravity they're used to on Earth. Much higher levels of g-force can harm the human body, and around 9 Gs most people will begin to black out as blood struggles to reach the brain. To combat risk, astronauts wear custom g-force clothing and train their bodies to better tolerate high-G pressures.

Negative Gs, or "eyeballs out" acceleration forces, produce the exact opposite sensation, where blood rushes toward the head as an aircraft accelerates downward faster than normal freefall. The danger zone is narrower for negative G-forces, and confusion and loss of consciousness can occur closer to -4 or -5 Gs. And then there's zero-acceleration force, or zero G, the astronaut sweet spot, which occurs when an aircraft or spacecraft is falling at the natural speed of freefall, creating a brief near-weightless environment.

Dressed in tight-fitting anti-G pants (or as I call them, "space pants") designed to help me withstand high levels of acceleration force, I hopped into an Extra EA-300 aerobatic two-seater and took off. The goal was to practice some of the techniques we had learned in the classroom to brace against high-G forces. The most effective was said to be an anti-G training trick called the Hook Maneuver, an onomatopoeic rendering of a respiratory trick involving the sharp inhalation "HOO" sound followed by exhalation "KEH." Over the clear skies of Daytona Beach, Florida, I HOO'd and KEH'd until the peripheral fields of my vision started to grey, and as we pulled 5.5 Gs, I experienced forces equal to nearly six times my body weight. I hadn't even finished my KEH exhalation before we rolled upside down and leveled out, rapidly dropping to zero Gs. It was exhilarating. Over and over I strengthened my tolerance of positive and negative G-forces, always ending with the short-lived euphoria of weightlessness.

Back on the ground, our crew headed back to Embry Riddle Aeronautical University, where the next component of our training paired us with spacesuit technicians to master donning and doffing our pressure suits, learning to self-pressurize to +3.5 psi, and venting pressure as necessary. Once fully pressurized and comfortable operating our own life-support equipment, we took turns climbing into a spacecraft simulator, where we simulated a fully suited suborbital science mission. The simulator was programmed with a Virgin Galactic flight profile, and our classroom training had prepared us to operate the camera instrumentation and complete our science objectives during suborbital flight: building unprecedented models of our upper atmosphere through tomographic imaging and *in situ* sampling of noctilucent clouds.

To successfully complete the program, we needed to execute a perfect run in the spacecraft simulator, deploying our camera instrumentation right before the spacecraft hit apogee, and imaging the elusive noctilucent clouds from space before retracting the probe for reentry. It takes a special blend of precision and patience to operate small instrumentation in a spacesuit, even with gloves designed for optimal dexterity. By the time we graduated, our candidate class was more than a crew; we were a space family that had been trained to support each other in the most extreme of environments.

The diversity of our crew was staggering. With four men and four women, we had perfect gender parity, something I wasn't used to seeing in the aerospace industry. The international diversity was also inspiring. As an American, I was accustomed to being in the majority nationality for space programs, but my crewmates had traveled from all over the world to follow a shared dream of spaceflight and prove that they had the right stuff to contribute. In fact, many of my crewmates came from countries with no national space programs of their own. I was surrounded by fighter pilots and trauma surgeons and materials scientists and a range of other impressive backgrounds. After working together as crew and successfully completing our training program, we had each earned the dream title of scientist-astronaut candidate and taken one giant leap closer to realizing our goals.

After graduation and the awarding of certificates, I posed shoulder to shoulder with my seven fellow trainees. Staring back at the image and the colorful assortment of various national flag patches, I remember thinking, "This is the future of spaceflight." Suborbital spaceflight is the great equalizer, democratizing access to space for teams of scientists around the world like ours. It's an incredible feeling to know my dream of spaceflight is not an "if" but a "when." In the meantime, my training qualified me to conduct a number of exciting space-research missions right here on Earth, where I've enjoyed testing spacesuits and flying in microgravity.

My role as a scientist-astronaut candidate with Project PoSSUM and IIAS opened up a whole new world of research and citizen science to me. While I still held a passion for the mesosphere, I also fell in love with bioastronautics, a field of research focused on human factors and the biological effects of spaceflight. Like the upper atmosphere, the effects of microgravity aren't particularly well understood, largely due to our limited ability to access the environment. The only way to evaluate technology and human performance in space is to conduct research in microgravity, or at least a high-fidelity simulation of it. Terrestrial analog habitats have their limits; to truly experience the weightless sensation of spaceflight you need to take to the skies, and four years after a crew rotation at the Mars Desert Research Station, I joined a test-flight crew in a specially modified Falcon-20 experimental aircraft to do exactly that.

In partnership with the Canadian National Research Council (CNRC), I've flown onboard a series of microgravity research flights as both a payload specialist and human test subject. To simulate microgravity and create a near-weightless research environment, an aircraft plans a parabolic flight profile not unlike a roller coaster. You can imagine an aircraft pulling up and delivering those "eyeballs in" positive G-forces, leveling out briefly at the top of the parabola, and then pushing gently over the top of that curve, easing into a twenty-five- or thirty-second-long freefall where everything in the aircraft is weightless. This arc is repeated over and over again throughout the flight, allowing researchers and their experiments precious minutes in the environment of microgravity. By adjusting the steepness of the parabola, pilots can also create reduced-gravity environments to mimic the sensation of lunar (one sixth) or Martian (one third) gravitational forces.

In the past, I had flown onboard Zero Gravity Corporation's modified Boeing 727 (nicknamed G-Force One), so I was already familiar with the sensation of floating freely throughout a padded cabin, where I could fly like Superman from cockpit to tail, catch individual Skittles or water droplets on the tip of my tongue, and perform a series of acrobatic moves that would be physically impossible for me to replicate on Earth.

The Canadian National Research Council's modified Falcon-20 is a great deal smaller, but its cabin much more accurately simulates the interior of a commercial suborbital spacecraft, and the sensation is equally euphoric. Those same parabolic flight profiles have allowed me to test a number of experiments in microgravity, ranging from biometric analysis and spacesuit performance to fluid configuration and solid-body rotation. Over the years, our team has collected valuable data for a number of research collaborators, including the Canadian Space Agency, Final Frontier Design, and the Massachusetts Institute of Technology.

To experience weightlessness, free from the effects of gravity, is a peculiar feeling. Imagine floating in the pool, limbs suspended and relaxed. Subtract the sensation of water beneath you and you'll have a rough idea of what you could expect to feel in microgravity. For those with sensitive stomachs, though, it can be much less serene. While the downward curve doesn't produce the same stomach drop as a rollercoaster, NASA astronaut candidates dubbed their training aircraft the "Vomit Comet," a reference to the disorientation and nausea some folks experience, also known as space motion sickness. While I've never experienced anything but joy in microgravity, I've certainly been on research flights with sick crewmates, and it seems incredibly unpleasant. A low-tech solution prevents the detritus from spewing across the cabin; Velcroed to the wall near everyone's seat is a small paper bag not unlike one you might find in the seatback pocket of a commercial aircraft.

Aside from the sick bags, Velcro is the second most important carry-on item for parabolic flight. Even if you experience no disorientation at all, it can be a challenge to adapt to microgravity. In normal gravity, we're used to putting something down and knowing intuitively that it will stay put exactly where we left it. In zero G, that's not the case. Even a simple task like jotting down a number or measurement becomes a methodical and multi-step process in microgravity. You have to unclip a pen from your flight-suit pocket or tethered location on a wall and then reinsert it or resecure it immediately after use. Everything needs to be fixed in place or else you'll be dodging debris all over the cabin.

I could revel all day in the thrill of zero G, but I have work to do when I fly, and those responsibilities require intense focus and concentration. When serving as the suited test subject, I fly fully pressurized in a Final Frontier Design IVA spacesuit, practicing ingress and egress from my seat in microgravity, evaluating glove dexterity, and operating a number of payload experiments from various research institutions. Underneath my spacesuit, I've also tested an exciting experiment from the Canadian Space Agency: the Bio-Monitor undershirt was a comfortable, wearable smart shirt designed to fit into an astronaut's daily routine and provide real-time tracking of vitals like blood pressure, temperature, breathing and heart rates, and blood oxygen levels. There's a lot we don't know about the long-term effects of microgravity on the human body, but we do know it takes a significant toll. Astronauts maintain a rigorous exercise program in space to combat the accelerated muscle atrophy and bone density loss, and careful monitoring of an astronaut's vitals through integrated technology like the Bio-Monitor can help provide valuable insight for medical and research teams on the ground.

My flight crew was asked to help support data collection and validation of the Bio-Monitor ahead of the experiment's launch to the International Space Station, which meant our team had the chance to test it out in microgravity while completing a series of exercise maneuvers in zero G. For me, this meant untethering from my seat in microgravity, navigating to a pair of footholds on the floor of the cabin, and then performing a repetition of squats and other aerobic maneuvers. Even though you're free from the forces of gravity, exercising in a spacesuit is a lot more difficult than it looks. We collected a lot of data, and a few weeks later we had the satisfaction of watching the Bio-Monitor experiment launch to its new home on the International Space Station with Canadian astronaut David Saint-Jacques.

On another flight, I was a spacesuit assistant to the suited test subject, helping my suited crewmate go visor down, pressurize, and vent pressure as necessary. That flight also drove home the importance of our many hours invested in crew cross-training. When another crew member fell ill (not uncommon in zero G), I was able to jump in and take over as crew biomedical monitor, monitoring crew CO_2 levels and ensuring that our suited subject's core vitals, like heart rate, core body temperature, and pulse oximetry readings, never went out of safety bounds.

I also monitored vitals of my own—shortly before one flight, I swallowed a tiny experiment from the Canadian National Research Council, a pill-shaped Bluetooth device designed to track my visceral core body temperature in flight. Once digested, I was able to "connect" the pill in my stomach with a small handheld tablet to track my vitals in real time. For at least fifteen minutes I amused myself with pairing and unpairing my body to the device, congratulating myself on having been born in such an age when this kind of human-computer interface was possible. For the next few days, I was instructed to wear a bright-yellow medical bracelet warning emergency service personnel that I was temporarily ineligible for an MRI. Welcome to the future, folks.

Crew movements and experiment protocols are carefully planned out for each parabola, but the first two are almost always reserved to let the crew acclimatize and ease in to the strange and wonderful sensation of zero G. During these two parabolas, crew members often have an opportunity to float a small personal memento before the experiments. I always take that opportunity to fly my daughter's mission patch, a special keepsake designed right before her birth by the talented NASA mission-patch artist Tim Gagnon. Watching it float in front of my helmet is enough to dispel any imposter syndrome that manages to creep into my world; it's a small but tangible reminder that there's no one specific background or degree required to contribute to space exploration. There is room for all of us in space, whether professionally or recreationally, and access to those opportunities will continue to widen for my daughter's generation.

A century ago, airplane flight was a novelty reserved mostly for military test pilots. All early flights, by the very nature of operating absent regulation, were experimental. It wasn't until the 1920s that commercial aviation soared and the private industry stabilized through standards, support infrastructure, and public promotion. Today, airline travel has integrated into society as a critical component of our everyday lives; passengers around the globe take more than four billion flights per year. It's easy to imagine a world a century from now, when the novelty of spaceflight has experienced a similar trajectory, changing the course of how we live, work, travel, and explore.

Currently we find ourselves at the doorstep to the golden age of spaceflight. The technology has arrived, but practical application remains mostly limited to government, military, and experimental crews. Space tourism, and suborbital spaceflight in particular, is about to change that, opening up the cosmos for civilians, students, scientists, and tourists alike. To deride the value of recreational spaceflight would be shortsighted in the face of humanity's ultimate astronomical trajectory. Worse, it's a failure of imagination.

Space tourism represents an opportunity to commercialize low Earth orbit, and while the earliest ticket prices may well be prohibitively expensive for some, the fact that any flight opportunity exists as an option for the general public is an important milestone, and one that ultimately benefits society as whole. Baked into a suborbital tourism business plan is an engineering and sales challenge to enable as many people as possible to experience the wonders of spaceflight. Tossed out from the formula are many of the restrictive requirements that have disqualified the dreams of so many astronaut hopefuls in the past. For suborbital spaceflight, you don't need 20/20 vision, perfect hearing, or a height between sixty-two and seventy-five inches. And if your conditions are well-controlled, even pre-existing medical issues aren't necessarily a deal breaker for safely undertaking spaceflight.

In 2017 the Federal Aviation Administration contributed funding for a study that demonstrated the ability of most members of the ordinary public to withstand the stress of suborbital spaceflight. To test their fitness for space, a group of volunteers between the ages of nineteen and eighty-nine climbed into a centrifuge at the National Aerospace Training and Research Center, where a gondola spun between two axes to generate up to 6 Gs of pressure. Virgin Galactic had designed their spaceflight operation's medical program with a diverse health group in mind from the beginning; in targeting a demographic who could afford the ticket price, they realized their passengers were likely to be a bit older and have lengthier and more diverse medical histories than your average astronaut.

An incentive to accommodate individuals who would have been deemed ineligible for early NASA missions represents a step-order change in qualification and access, one certain to broaden the future of bioastronautics, space medicine, and specialized equipment. For a species who will eventually need to find a new planetary home, we should be championing any business case that endeavors to democratize access to space for the broadest possible swaths of the human population.

The luxury price tag is also sure to attract companies looking to align their brands with aspirational experiences. The space industry can expect an influx of corporate advertising dollars, along with commercials, sweepstakes, and sponsored promotions, all contributing to a higher flight tempo and more robust customer market that will help lower the cost of operations. Eventually, suborbital spaceflight tickets will be accessible to civilians at a price point similar to premium airline tickets.

A robust suborbital spaceflight market will also help mature the development of other (and to some, more practical) travel capabilities, like point-to-point hypersonic flights that could ferry passengers between New York and Tokyo within two or three hours. Just as aviation first enabled us to reach other states or countries in a fraction of the previous travel time, so too will point-to-point suborbital spaceflight technology unlock even more rapid travel. Journeys that currently take hours will eventually take minutes, once again reinventing global economies and societies.

Perhaps most exciting, suborbital space tourism approaches spaceflight with an entirely new objective. For the first few hundred humans to venture into space, the flights were focused entirely on function. But for the next few hundred humans to travel to space, we have the opportunity to optimize the experience. The spacecraft interiors are designed to maximize the sense of wonder for floating passengers, boasting large cabin windows from which space travelers can gaze down at their home planet and all 7.7 billion fellow members of the species. This time, the incredible experience isn't just a by-product—it's the main mission. The next wave of space travelers won't all be engineers, and that's entirely the point.

CHAPTER 10

THEY SHOULD SEND
POETS

THE LAUNCH OF SPUTNIK MAY HAVE KICKED OFF THE SPACE AGE, BUT THE APOLLO MOON landings truly cemented it, unlocking a critical milestone in humanity's astronomical trajectory. The treasure trove of scientific data returned from our earliest journeys beyond low Earth orbit more than justified the missions' enormous effort and expense, but for a sentimental species the data alone could never convey the passion—and the profound impact—of that journey.

The soil mechanics experiment wasn't the thing that moved the public, despite the very important investigation into the properties of lunar soil. Similarly, you would be hard-pressed to find a single member of the space-crazed public in 1969 (or today) who would cite the laser-ranging retroreflector or the passive seismic experiment among their lasting impressions of Apollo 11. The data collected from lunar surface experiments represented incredible and important feats of science, but that wasn't what gripped hearts and imaginations around the world.

Instead, humanity's collective memories and lasting impressions of Apollo were shaped by people who looked at the engineering genius of Apollo with an artist's eye. The audio feed delivering the crackle of Neil Armstrong's voice a world away, the video footage broadcasting humanity's first few tentative steps into a mysterious new environment, an instantly iconic photograph of a footprint on another celestial body, Buzz Aldrin's poetic and all too human descriptions of solitude and desolation, and stark images of an American flag that evoked both triumph and humility.

The crew of *Apollo 11* returned home to a spectacle of parades, photographs, and interviews from an inquiring public who were interested less in the composition of lunar soil and more in how it *felt* beneath their moon boots. In the decades that followed, inspired civilians produced an avalanche of books, films, songs, and artwork that paid homage to humanity's greatest achievement. In turn, those creative works inspired new generations of engineers and scientists and motivated them to develop the skills necessary to push further our capabilities in space exploration. The cycle continues, with art and science representing the eternal yin and yang of human advancement.

Through that lens, the promotion of science, technology, engineering, and mathematics alone begins to feel incomplete. These STEM fields prioritize discovery, data, and problem-solving, but their focus begs the question of how we begin to identify and prioritize the problems we want to solve in the first place. A more well-rounded curriculum is the promotion of STEAM, introducing art into the acronym, a discipline that prioritizes passion and creativity—the essential building blocks of innovation in any industry. Just like any other skill worth developing, creative thinking requires practice and intentional nurturing.

Early STEAM exposure and curriculums are investments in students as both individuals and as the future of society. Alongside every rocket scientist and engineer constructing a spacecraft to carry the species further into the cosmos, we'll need experts considering the far-reaching implications of that journey—those versed in the studies of humanities, philosophy, history, anthropology, ethics, law, politics, religion, or art. Scientific advancement doesn't exist independent of human impact; it requires a societal discussion of what we need to invent and why, or sometimes whether we should invent something at all. In pairing the arts and sciences together, we create a Möbius strip of creativity and unlock the deepest reservoirs of human potential.

It's exciting to imagine the profound societal impact of a species beginning to collectively look upward. For the next generation of space travelers, how might a new perspective on our home planet influence the ways in which we approach our future? From the firsthand accounts of astronauts, we know a space-induced cognitive shift in awareness occurs. The phenomenon was named by author Frank White as the overview effect, and it has long been reported by those who have had the unique opportunity to view the Earth from outer space. In each instance, an astronaut's journey to space has led to a rediscovery of their home planet.

For Apollo 14 astronaut Edgar Mitchell, the overview effect produced an epiphany of global unity and an immediate distaste for the fractured geopolitics back home. "You develop an instant global consciousness, a people orientation, an intense dissatisfaction with the state of the world, and a compulsion to do something about it. From out there on the Moon, international politics look so petty. You want to grab a politician by the scruff of the neck and drag him a quarter of a million miles out and say, 'Look at that, you son of a bitch.' "

For my former boss, NASA astronaut and ISS Commander Mike L-A, the overview effect produced a feeling of unity. "I've never had another experience like it. Seeing Earth from above shifts your perspective from a narrow, geocentric view to a global and almost cosmic view. We're all in this together."

And what will it look like to send hundreds of people to space each year, rather than handfuls? The optimism of the overview effect is perhaps best summed up by commercial astronaut Beth Moses, who publicly reflected on the implications after her suborbital spaceflight on Virgin Galactic's *SpaceShipTwo*: "I believe if a greater slice of humanity can experience spaceflight, it will translate to untold benefits and changes on Earth. What if every world leader saw Earth from space? It might be a gentler, kinder planet."

We have plenty of work ahead of us to open up the final frontier to broad swaths of the human population, but it is certainly a future worth working toward. In the meantime, as we democratize access to space, it's equally important that we continue to champion and contextualize the journey for those around us, sharing the triumphs, the failures, and everything in between. We are a species of storytellers. From sitting beneath the stars around the earliest fire, to the crowds gathered beneath the flame of a rocket headed for another planet, our species has always motivated ourselves forward through story and shared perspective.

Once I discovered my passion for space exploration, sharing that passion with the public came easily to me. In some ways, my lack of a traditional engineering degree amplified my space communication skills; I naturally gravitated away from nitty-gritty technical concepts in favor of the big picture takeaways that had gripped me in the first place, easily distilling the themes that resonated most strongly with the general public.

As I set out to nurture my natural tendencies into actual talents, I found inspiration in an invaluable but often-overlooked niche of science professionals: the science communicators. Growing up, I had no idea that popularizing science was a viable career, even though I had plenty of exposure to talented folks operating squarely at the intersection of science and pop culture.

I came to understand that the secret sauce that elevated a brilliant scientist to a household name was their ability to connect with the general public. From Carl Sagan and Stephen Hawking to Bill Nye the Science Guy and Neil deGrasse Tyson, the most effective scientists are capable of communicating not only data and results but also human impact—the *story* of the data—to the public. Those science sorcerers who can wield data to spark curiosity, popularize a field, or inform public policy are all practicing the timeless art of science communication, or SciComm. In addition to entertaining and inspiring, science communication can also play an important role in increasing the public understanding of science—and scientific literacy is a cloak of protection for modern societies.

The ability to briefly capture the attention of the public is impressive, but to hold that attention over time is invaluable. Considering the spectacle of human spaceflight, one might imagine that astronauts would be professional celebrities, perhaps household names worthy of tabloid gossip and paparazzi shots. For better or worse, and despite being well-deserving of recognition, we know that's not the case. Beyond Neil Armstrong and Buzz Aldrin, who had the extraordinary fortune of being assigned to the historic Apollo 11 mission, many folks struggle to name even one of the ten moonwalkers who followed. Similarly, esteemed NASA astronaut Sally Ride permeated pop culture consciousness through novelty of gender rather than her proven skills; and despite the intense focus on Sally's personal life, the average American would be hard-pressed to name the second woman in space, or the first to command a space shuttle or perform an EVA.

Of course, a lack of name or mission recognition in no way diminishes the impressive achievements of NASA's spacefaring pioneers, and neither should it be construed as a critique of the public's appreciation for such achievements. For fifty years NASA has (understandably) prioritized sending engineers and scientists to space, and aside from the tragic case of Christa McAuliffe, public engagement was rarely the mission priority.

And yet space is ubiquitous—no more than sixty miles away from every single person on Earth. So how to bring space down to Earthlings in a way that captures hearts and imaginations? If we want an emotionally invested public, we must bridge the gap between science and society, the ultimate quest of science communicators.

The impact of an engineering feat becomes far wider reaching if the engineer is able to translate advanced technological concepts into terms that allow the general public to share in the triumph. Similarly, a mathematical model finds real momentum when the mathematician is able to extract societal impact from the integers. But a technical degree is by no means prerequisite to spreading enthusiasm for science or space among the public— these days, increasing awareness and excitement around scientific innovation can be accomplished by anyone with an internet connection. And for those who wish to formalize a career around it, there are a number of nontechnical roles in the space industry that are underpinned by strong communication skills.

Take space reporters, professionally trained in journalism and capable of leveraging that expertise to share the happenings of the space industry with the general public, contextualizing technological progress through a wider societal lens. Similarly, public affairs and media positions provide a platform for public engagement on behalf of government agencies or commercial companies working to advance space exploration. In a world of livestreamed launches, webcast press events, and the emergent field of space tourism, communication skills have become mission critical in the space industry.

While we wait for spaceflight to be experienced and translated through the artistry of poets, painters, musicians, and so many other talented civilians who will soon journey beyond our home planet, there already exists an army of space and science communicators right here on Earth, well equipped to share the what and the why of the final frontier, bringing space down to Earth for us all. Following in the footsteps of those science communicators who had first sparked my own love for space, I aimed to share the wonder of human spaceflight with the masses. I was neither a poet nor a brilliant engineer, but I was an earnest space advocate, an active contributor to scientific research, and a natural communicator. In spreading awareness of the future we could collectively unlock, I dreamed of motivating people to make the most of life in a Space Age. Luckily for me, reaching thousands of people became a whole lot easier with the birth of social media.

I was fifteen the first time a piece of my content went viral on the internet. It was Christmas morning in 2006 and I filmed a low-resolution video of my father opening his present. "What'd you get for Christmas, Dad?" I shouted off camera. A natural entertainer, my dad made a production of slowly peeling back the wrapping paper before jumping up and down in his bathrobe, pumping his fist in the air, and screaming, "Yes! YES! XBOX 360! YES, yesssss!"

YouTube was just one year old, and without thinking too much about it, I uploaded the clip to the new video-sharing platform with the title "Crazy Xbox 360 Dad" and went back to my family's festivities. Nearly a quarter of a million views later, my mind was blown, and I learned a few important lessons about the emerging world of social media.

First, it amazed me that the bar to going viral could be so low. Creating a quick laugh for the collective internet had cost me no money, required no special equipment, had taken barely any time, and required a negligible amount of mental energy. That this shaky cell phone clip lasting all of eighteen seconds could reach a quarter of a million people was both exhilarating and disconcerting. If I had any idea how many eyes would be on this, I might have put a little more effort into the production.

I also learned that those eyeballs were valuable. I knew nothing about digital advertising, but soon after the video took off, YouTube sent me an offer to enroll in something called revenue sharing, an arrangement in which ads would automatically run at the beginning of my video, and my share of that advertising revenue would be paid out in monthly installments. That anyone would willingly hitch their brand to footage of my father hopping around in a bathrobe—and *pay for it*, no less—astonished me. The footage was even licensed by an electronics company for a commercial featuring a mashup of similarly enthusiastic reactions to video game gifts.

The money generated from the video was insubstantial, but I experienced the equivalent of a gambling rush watching the cents add up to dollars from the faceless viewers who had each sacrificed thirty seconds of their lives to watch a short commercial followed by entertainment in the form of my dad's jubilation. That opened my eyes to another essential internet truth: social media is absolutely addictive. My father was enjoying the whirlwind just as much as I was—from our respective corners of the house, we'd sit refreshing the YouTube page over and over, watching the views climb up and shouting out when we hit a milestone. One hundred views! A thousand!

Ten thousand, fifty thousand, one hundred thousand! My face wasn't in a single frame and I already felt a powerful wave of secondhand celebrity.

Looking back, it was a blessing that only my voice was featured in that first brush with internet notoriety. I needed that layer of distance to fully digest and appreciate my final and most important social media lesson: the internet can be an ugly place and the comments section is lawless. In the beginning, my notification settings had YouTube send me an email announcing each new comment; as the reactions reached the hundreds, I figured out how to turn that feature off, but not before confronting dozens of unfiltered reactions in my inbox. Thankfully, most were positive, with a majority of people laughing and tagging friends who would find the clip equally hilarious. Others made jokes at the expense of Xbox and how disappointed my dad would be once he discovered the glitches that plagued the new console. Not even the mocking comments directed at my dad bothered me; I laughed out loud at a small but insistent group of commenters shrieking about what they thought they saw as his bathrobe shifted in the jumping.

It hadn't occurred to me that anyone could possibly have a comment about me, the phantom narrator; somehow, that felt like a step too far. But of course, nothing is off limits in the comments section. "She sounds like Meg Griffin from *Family Guy*," announced one comment, which attracted at least nine "likes" in agreement. "Shut up Meg," agreed someone else. My fifteen-year-old brain panicked about the revelation (to be fair, the observation wasn't wrong) and what the Meg association could possibly imply about me. "She sounds hot" rebutted someone else. I immediately clicked like on that comment, silently willing others to do the same. A handful of really vulgar themes followed, all of which I deleted immediately, humiliated by the thought of anyone else reading them. It suddenly felt invasive that people could so freely discuss me and my family without knowing us at all. But a family scene from my living room had made its way into the living rooms of hundreds of thousands of other people, and each of them felt entitled to share their reactions—the good, bad, and the downright awful.

Despite the obvious risks of sharing content with the world, I was gripped by the power and potential of the internet, and my media savvy increased with each new platform. That same year, Twitter launched, and then Facebook opened itself to the public, followed a few years later by Instagram and then TikTok. And so on. My social media debut had been a silly YouTube video, but I gained a deep appreciation of the power and potential these platforms provided for modern storytelling, thought shaping, and perspective sharing.

The early exposure to unexpected attention ensured I remained intentional about what I shared with the digital world going forward, and it gave me a head start on developing the thick skin I would eventually need for building a large, lasting, and hopefully inspirational brand around my passion for space.

When I first began sharing my journey into the space industry, I had the digital support of a handful of friends and followers.

My excitement proved infectious; over the next few years, I picked up a couple hundred thousand more. For the greater part of a decade, I've openly shared vignettes from my life with followers all over the world. In return for that transparency, I've received support and encouragement through the highs and lows of personal and professional growth, research, a rising media profile, new business ventures, spaceflight training, microgravity campaigns, marriage, and parenthood. Even better, I've connected with a large community of like-minded space enthusiasts and have had the privilege to introduce the industry I love to thousands of others. Together, we've created a community where we can celebrate spaceflight and redefine "the right stuff." Best of all, the communication is two way; through comments, reactions, shared perspectives, and personal anecdotes, my life has been greatly enriched by people I've never met in person, but who I've come to know virtually.

In addition to oversharing on the internet, I attribute much of my (or anyone's) personal brand growth to relatability. Starting out, I was a fellow outsider to an exciting industry, and my well-documented attempts to break in were captioned with exclamation points and uninhibited awe. Of course, the mission to wedge my foot in the door was filled with a number of pinch-me moments, each of which I dutifully captured, cropped, captioned, and uploaded to social media. Scrolling through my Instagram feed, you'll find a photo mosaic of my rise from wannabe to can't-believe.

The other thing I had going for me was an unintentional penchant for stereotype busting. When you hear "aerospace and defense," I'm not necessarily the first person you'd imagine. It turns out there are a lot of people who appreciate seeing someone in my career position who is approachable, excitable, and whose hair is sometimes bright pink. My very visible existence stood out as a challenge to the preconceived ideas of who we picture when we imagine a scientist or an astronaut, and I made a point to double down on those themes to prove that there's a place for everyone in science and space exploration.

There are plenty of obvious benefits that come along with a giant social media platform generating millions of views, and I might be biased here, but I've never found the word *influencer* to be pejorative. The content I create has always been explicitly designed to influence the way people think about their place in the Space Age, and to inspire folks to get involved in humanity's next giant leap. And while the goal should never be simply to attract a large following, a strong brand presence and digital name recognition can open up a number of exciting opportunities along the way.

For anyone endeavoring to build a SciComm brand of your own, having a cohesive and curated internet presence will help ensure the right opportunities find you. I stumbled my way to internet space fame, but after watching a number of friends intentionally build digital empires from scratch, I can share five main pieces of advice for growing a platform to share your passion with the world.

HOW TO WIN FOLLOWERS AND INFLUENCE PEOPLE (SCICOMM EDITION!)

1. Lay the Groundwork

To ensure as many internet roads as possible lead back to your body of work or field of research, you'll want to expand your internet presence beyond your social media platforms. Start small with a blog or website (ideally FirstNameLastName.com) where you can host a headshot, a short biography or personal mission statement, and a portfolio of work or reel of relevant content. Remember that a headshot doesn't always have to be a shoulders-up portrait; it can also be an action shot if that more accurately captures what you're all about. Ideally, you can use that same profile image across all social media platforms, because eventually you'll be using that same photo for speaking engagements as an esteemed voice in your field and the consistency will help new fans find you. (Manifest it, baby!)

2. Start Communicating

After setting up your personal website and synchronizing your social media accounts, start publishing reflective essays or opinion pieces on a personal blog or writing platform like *Medium*. (You'll also be able to include links to these pieces as writing samples when you're pitching popular websites or publications.) It's also helpful to decide on your personal SciComm mission: do you intend for your content to educate, inspire, or create awareness? In my case, my specific goal starting out on social media was to increase awareness about the progress of the commercial spaceflight industry. I would pull from both headlines and archives to share interesting, informative space news or history with my followers along with my own editorialized twist. Later on, as my own space career and profile progressed, my page became more inspirational, and I switched to sharing photos of my daily life as a space professional in that industry to hammer home that this future belongs to each one of us.

3. Carve Out the Time

As the old adage goes, if it's worth doing, it's worth doing well. If you're communicating science through a visual medium, you'll want to invest time and thought to compose your image or video and craft the message you believe will most strongly resonate with the people you're trying to educate or inspire. As your efforts start paying off and your following grows (and it will!), don't take any hard-won engagement for granted. I've always made an effort (and still do) to respond to each comment and message and to actively engage with folks who are investing time with me. Always remember that your content is reaching real people. Ask questions, engage in dialogue, and invite new perspectives.

4. Invest in Your Own Success

I've seen folks leverage their digital platforms to bring them closer to their real-life career goals. Social media means most people, famous or not, are only a few clicks away, so you could try a digital cold call to some role models in your field to see if they might be open to doing a short interview on your platform. Similarly, you can reach out to companies or organizations who are particularly well aligned with your subject area to see if they might be open to an educational collaboration. And while you're investing in yourself, consider investing in your platform itself: a few high-quality headshots, a website makeover, or your own personal photography or lighting equipment can help bring a little polish to your digital profile.

5. Embrace Your Multitudes

When I think about the science communicators I follow most loyally online, it strikes me that I know a lot of small, intimate details about the lives and careers of people I don't know in real life. I'm not just following them; I'm emotionally *invested* in them. Like anyone else, I've evolved over time as a person. Sprinkling into my science content personal moments unrelated to my career in space (escapades with my toddler, travel highlights, political frustrations, social activism) has cost me followers in some cases, but it has built a much stronger connection between myself and the folks with whom I really want to be engaging. Sharing your personality helps followers better understand your multitudes and your unique perspectives.

If I may offer one last bonus piece of advice (which I can, it's my book) it would be to go all in and embrace the chaos. Once you've built a platform and committed to putting yourself out there, be bold and try new things. For me, that meant trying my hand at fashion design, a long-time alternate dream career. Working in the space industry has been every bit as exciting as it sounds, but it hasn't necessarily been glamorous. I remember trying often to tone down my look to fit in and feeling frivolous if I was the only one clacking down the hallway in heels amidst an army of suits. Somewhere along the way, I gained confidence and realized my interests in science and fashion weren't mutually exclusive.

After exhausting my own closet's options for screaming "SPACE!" with my ensemble, I decided to source fabrics and materials and create some samples of the space-themed wardrobe of my dreams. I enjoyed the fashion design process so much that I ended up creating and producing an entire limited-run collection in 2016 called "Paper Rocket."

The label didn't last, but I shared the entire process with my followers. My small batch of inventory sold out immediately and showed me how many other women were eager to find high-quality items that proudly displayed their interests. It was an expensive experiment for me, but I learned a ton and briefly brought to life some awesome space fashion. Importantly, my samples were all sized to my exact measurements, so despite the money pit, not at all was lost: I ended up with a closet full of perfectly tailored space clothes that I would have happily paid double for. A few years later, I revived Paper Rocket in the form of a space fashion and lifestyle website where I continue to share a different side of my space research—the kind where I share and support my favorite space fashion and accessory designers that I've discovered over the years of scouring the internet.

Of course, my favorite space-themed outfit will always be a spacesuit. I've been fortunate enough to test Final Frontier Design spacesuits in a number of extreme environments, from microgravity flights, to a rotation at the Mars Desert Research

Station, and even on the red carpet (best dressed for sure). I always feel like the best version of myself when I'm in a spacesuit and I never forget what a privilege it is to suit up and do science. That said, work becomes much more personal after having children. It's inspiring to think that the problems I have the privilege to work on today will impact my daughter's future, but the balance can be incredibly overwhelming.

My daughter Delta was brand new to the world when I completed my scientist-astronaut candidate program. In the months and years that followed, I was away multiple times for varying lengths of time, and I missed a number of important "firsts" in her little life. I always miss her when I travel, and it can be difficult to push away feelings of guilt. When it gets me down, I like to rewatch an old interview of myself delivering an emotional but genuine answer to the question of how I coped with being away from my infant so much: "I just take it one trip at a time, and I always remember that investments of my time are investments in her future." (Excellent point, postpartum Kellie.)

The message resonated with so many of my fans because it's clear that it applies far beyond space endeavors. It applies to all of the moms and dads who get up every day and do their part in all different fields to help engineer a brighter future for the next generation.

And no matter how much of a super mom you might be, adjusting to motherhood is a draining process. Working full time adds yet another challenging dimension. I was looking for the elusive "work-life balance" and the first few weeks back at work felt like absolute failure as I tried to balance competing responsibilities and find a new rhythm. (Spoiler alert: newborns are rhythmless. It's kind of their thing.) After an exasperated text message exchange, a friend sent me a link to an article in *Scientific American* called "The Special Challenges of Being Both a Mom and a Scientist," in which author Rebecca Calisi perfectly described the Schrödinger's cat of motherhood: the expectation to work as though parenthood does not exist and to parent as if work does not exist. I was still exasperated, but boy did I feel *seen*.

I've been lucky to have supportive colleagues committed to helping me transition back to work successfully, a family willing to pitch in to ensure that success, and my own special corner of the internet cheering me on and leaving kind comments beneath every sleep-deprived selfie. Despite the soft landing, "work-life balance" has never felt like a realistic goal to me. For one, the concept of two distinct spheres—separate but balanced—has always felt like an unrealistic and borderline irrational goal for someone who decided they aspired to be an astronaut, a popular-science communicator, a fashion designer, a wife, a mom, an author, and an involved member of the community. And for two, life *is also* work.

Instead, I've learned to embrace something a clever colleague coined *work-work balance*, and I would have saved myself a lot of angst if I had simply accepted it for what it was from the beginning. On the plus side, in the lawless reality where your work is personal and your personal life is work, I've had plenty of excuses to include my family in my space-capades. Over the course of years advocating for NASA's Commercial Crew and Cargo Program, I've had the honor to attend and celebrate a number of SpaceX milestone launches. Perhaps most exciting was the inaugural flight of *Falcon Heavy*, which launched in February 2018 from the legendary Apollo 11 pad at Cape Canaveral. I had just recently given birth, but I decided there was no way I was missing this flight, and I was doubly thrilled at the prospect of sharing this special moment in spaceflight history with my newborn daughter.

With the ability to propel the equivalent of a fully loaded 737 jetliner (119 thousand pounds) into orbit, the launch of *Falcon Heavy* was the debut of the world's most powerful operational rocket, doubling the payload capacity of the next biggest rocket at the time. The mighty rocket was designed to one day carry crew and supplies to deep-space destinations and the inaugural payload for this first test flight was precious cargo indeed: loaded into the center of the payload bay was SpaceX CEO Elon Musk's cherry-red Tesla Roadster, including a dummy astronaut passenger nicknamed Starman. Whoever had flying cars on their 2018 bingo card deserved a win by technicality.

Roughly ten weeks postpartum, I packed up the diaper bag and traveled to NASA's Kennedy Space Center with my daughter and my mother for the history-making launch. With "Feel the Heat" tickets, three generations of women in my family settled onto the grass outside of the Apollo Saturn V Center, less than three miles away from the launch pad, ready to hear what twenty-seven engines and more than five million pounds of thrust at liftoff sounds like. The spectacle wouldn't end with the launch;

after liftoff and the separation of the rocket stages, SpaceX would attempt to land all three of *Falcon Heavy*'s first-stage rocket boosters back on Earth. With infant ear protection in hand, we were ready to experience one launch, three landings, and six sonic booms.

The visuals were instantly iconic: Starman buckled into a cherry-red Tesla Roadster, leaving Earth's orbit on a billion-year cosmic road trip, drifting through space with a copy of *The Hitchhiker's Guide to the Galaxy* in the glove compartment, David Bowie's "Space Oddity" on the radio, a "Don't Panic" sign on the dashboard, and an engraved circuit board that says, simply and poignantly, "Made on Earth by humans."

As with all rocket launches, first you see it, then you hear it, and then you *feel* it. When the sound barrier broke over my daughter's head, it struck me that she had been born exactly at the beginning of a new era in space exploration. And just like their historic space station berthing back in 2012, the moment was greater than SpaceX's alone; it also represented the culmination of years of hard work from awe-inspiring companies, advocacy groups, regulatory bodies, and government agencies. So many individuals and organizations working together toward a shared dream of expanding humanity's footprint in the solar system—a dream my daughter's generation will certainly see realized. For me, space has always represented the best of humanity—the spirit of exploration, the genius of engineering, the quest for knowledge, and the indomitable hope that we can survive our time so that the next generation may live in the future. It's a cause that reminds me the work-work balance is a downright privilege, and a vision that motivates me to communicate that potential and positive energy with the world.

When I think of my best SciComm moments, I think of that day, as three generations of women in my family watched the world's most powerful rocket leave planet Earth. I livestreamed the launch and impressive technical details to my followers, but I also shared with them my raw emotion and awe of humanity, overwhelmed by the future awaiting the tiny baby in my arms. I remember telling everyone watching live that roughly 10,000 generations of women stand behind my baby daughter. That's 10,000 times that the baton of survival has been passed down through our species—through all those predators, mass extinctions, diseases, and wars—so that my little girl and the generations that come after her could live to see today, a Space Age. It's on all of us to make the most of the opportunity, and we'll be damned if we let the baton fall on our watch.

CHAPTER 11

THE BIG WHY

FROM OUR EARLIEST DAYS, WE'VE BEEN A
SPECIES OF EXPLORERS. FROM CRATERS TO
mountains, across oceans and new continents, we have slowly
but purposefully uncovered the mysteries of our home planet as
it hurtles through the vastness of space, each one of us a crew
member aboard Spaceship Earth. It's exciting to look back at
the past half century of achievements since our first tentative
journeys beyond Earth's atmosphere, but it's even more thrilling
to look forward. The next fifty years of advancement belong to
all of us—expanding our footprint in the cosmos is our shared
mission and our shared future.

The Apollo program took us further than our species had ever
been, some quarter of a million miles away from our home
planet. But we still have so far to go. The Milky Way galaxy,
which we have yet to probe the boundaries of, exists among
hundreds of billions of galaxies in the observable universe, the
very definition of potential. Of course, realizing that potential will
require big investment and bold commitment. We look toward
the sky with the certainty that answers to some of humanity's
oldest and most existential questions linger just beyond those
horizon lines. And we know that the impulse to reach for the
stars isn't a whim—it's a survival instinct. We find ourselves on
the cusp of the golden age of spaceflight, and how lucky that
we happen to be alive in this small, unprecedented window in
human history where interplanetary travel is finally possible.

In my career, I've trained for spaceflight, conducted research in
microgravity, and I've had the opportunity to work with rockets,
spacecraft, lunar landers, satellites, spaceports, spacesuits,

and more. Even better, I've been given the opportunity to share my experiences and reflections with the world. I built my career in an industry that until just a few decades ago existed only in science fiction like *Star Trek*. The writers in those shows imagined a future that probed the boundaries of human progress. That progress, the hope not only for survival but for prosperity, is why we continue to explore.

My professional journey is proof that there is no one specific degree or background required to contribute to the future of space exploration. Space is for everyone—it's our past and our future as a species. Just like Apollo, our next chapter will require the talents of artists, engineers, and everyone in between. Each one of us will be represented in humanity's next giant leap.

I feel so fortunate to have benefited from incredible mentors as I navigated a career around advancing humanity's footprint in the solar system. Richard and Laetitia Garriott de Cayeux, Stephen Hawking and his daughter Lucy Hawking, and Michael López-Alegría are just a few of my own personal role models, each of whom has demonstrated what it means to make the most of life in the Space Age. Over the span of my career they've shared with me their wisdom, their encouragement, and their passion for space exploration. Their guidance was critical when I first charted a course for the space industry, and in the interest of paying that forward, I'm delighted to share them with you.

As you read their own reflections, consider how far we've come since the first beep-beep-beep of Sputnik, how fortunate we are to have front-row seats to the final frontier, and how many wonderful discoveries await us and all those who will follow in our footsteps. Welcome to the Space Age—what a time to be alive.

A CONVERSATION WITH RICHARD GARRIOTT DE CAYEUX

What was it like growing up with an astronaut parent? Did you feel like space was in your blood? Did space travel feel "normalized" in your home?

I expect every child grows up believing the family and community environment they are in must be similar to what others experience. Not only was my father an astronaut, but my left neighbor Hoot Gibson was an astronaut and my right neighbor Joe Engle was also an astronaut. All NASA astronauts live near and train at the Johnson Space Center outside of Houston. The NASA facility and neighborhood was literally created out of drained swamp land, and thus not only did I live close to all of the astronauts from Mercury, Gemini, and Apollo, but those neighbors who were not astronauts were most always NASA contractors still deeply involved in sending humans to space. It wasn't until I went off to college that I met "The Sesame Street People," you know, butchers, bakers, firemen, and all the other jobs that make up "normal" life. So yes, space travel was not something you had to decide to do, it seemed like all of humanity were destined to fly into space...and soon!

Can you share a little bit about your career before becoming an astronaut? What were you most passionate about and how did that eventually translate to your dream of spaceflight?

About the age of thirteen, the NASA doctor, who was our family doctor, told me that because I was going to need glasses, I was no longer eligible to be a NASA astronaut. I was crushed: I had just been kicked out of the club that every adult I knew was in, and long before I even had the chance to decide for myself. After a few days of grief, I devoted myself to creating a commercial path to space, so that I could go! At the age of thirteen, I did not do much about that. However, just a couple years later, I found my first vocation: computer games. I fell in love with computers just at the dawn of PCs. My passion allowed me to become one of the first and most successful computer game developers. As money came in, I devoted the majority of my profits to opening commercial spaceflight! For many years, I invested in astronauts peeling off from NASA to start something entrepreneurial. Sadly, these didn't generally do well; I would argue that being a great astronaut does not necessarily make you a great entrepreneur. But eventually, I joined up with like-minded folks such as Peter Diamandis, Eric Anderson, Mike McDowell, and others, and combined we created things such as the X Prize, Zero G Corporation, Space Adventures, and more. Plus, I invested in companies like SpaceHab, XCOR, and others. When the X Prize was won in 2004, commercial spaceflight went from laughable to obvious. And the commercial space race began! Finally, in 2008, I was able to take my own trip to space aboard *Soyuz TMA 13* for a two-week stay aboard the ISS. But commercial spaceflight continues to build; we invested in SpaceX, and other members of the new space era. It's only just getting going!

After all of your early exposure and training, what about your spaceflight experience did you find most surprising?

I got in the best physical shape I had been in since high school in preparation for my training. Then, when I arrived in Star City, I discovered I was in way better shape than most all the others. It's been downhill ever since! It was also interesting to see how much fun, and not particularly hard, the training would be. For example, if you scuba dive, you already understand the life-support issues on station, associated with partial pressures of gases. If you can go online to get a ham radio license you can operate the spacecraft radios. It was all great fun, and all could be mastered by most of the educated general public.

Did you experience a cognitive shift after visiting space yourself? How has that overview effect impacted your outlook on life?

Absolutely. In fact, that was the most impactful part of the whole experience. I have regularly stated since that if we could get 0.5 percent of humanity to have this experience, the world would be forever changed for the better!

Imagine, you ride a rocket for 8.5 minutes to reach an altitude of only 250 miles, but a velocity of 17,210 mph. That's so fast that you circle the Earth in ninety minutes, you see a sunrise or sunset every forty-five minutes, and you cross whole continents in about twenty minutes. From only 250 miles up, you see great detail in the world below. You see the edges of tectonic plates, you see silt washing out to the oceans, smoke from fires covering whole states, haze from pollution, the burning of the rain forests, terraforming such as the Palm Islands in Dubai. You are captivated by what you see, and it feels like a fire hose of *truth* is pouring into your mind just from watching out the window. Finally, when I passed over Austin, Texas, where I lived, and saw the lakes, roads, and rivers I knew so well, I could even see the shiny dome atop my own home. I could also see most of Texas, which I have driven, biked, and camped across. Suddenly I had a physical shudder as I realized the true scale of the Earth by direct observation! It felt like watching a scary movie where the actor (Earth) stays the same size but the reality around them shrinks down rapidly. I still get goosebumps when I describe this.

Since then, I have felt much closer to the Earth, I have felt the importance of being a good steward.

If not a video game developer or astronaut, what do you think might have been an alternate dream career for you?

As an explorer and a creator, I'm in the best combination that exists, and I'd be hard-pressed to find another...but it would involve those two aspects.

What unique strengths and perspectives do you think non-engineers can bring to the space industry?

Jack Lousma, who flew with my father on Skylab, wrote me a great note when he heard I was flying. He noted that it was very important that people like me fly to space. He noted that he and my father were hired to do a specific job, not to explore other possibilities and not to communicate to others about what they had done. He knew that not only would I help find ways for humanity to thrive beyond Earth in ways that his hard-core engineering might not be able to, but it would be especially valuable when I got back to share the joys, challenges, and importance of spaceflight with others. I believe he was right.

Are there any early indications that either (or both!) of your children might be interested in becoming the world's first third-generation astronaut? What do they find most interesting about space and your experiences?

My children Kinga (eight) and Ronin (six) are just becoming old enough to really understand what space is and why going there is so difficult but rewarding. Their answer to "Do you want to go?" depends on which parent has spoken with them most recently. My wife convinces them that it's safer and more abundant here on Earth, and I, of course, convince them that exploring the vast unknown is the greatest adventure of all. By the time they are adults, I think space travel will be much more common, and at least far easier than it was for my first trip. I expect we will all fly together...more than once!

Your early investments helped kick-start an entire industry—what motivated you to get involved in the commercial space industry?

I knew that there would be no route for me, unless there was a route for many. I needed to help bring into existence a commercial route, so that I could go. I'd love to say it was more

"altruistic." But when it got down to it...I really wanted to go...
badly. It was worth risking every penny I earned in the games
industry, which I did, and it worked out great!

What aspect or milestone of humanity's future journey in space do you look most forward to?

Clearly Mars is the next major target. If Congress demands we
waste time and money on lesser pursuits, so be it. There is a lot
going on with NASA spending on jobs programs that is stuck in
old methods and ideas. But we are moving in the right direction,
however slowly.

Why does space matter? What's at the heart of the human quest and passion for the cosmos?

Space is essential to humanity *now* and for humanity to have *any*
future. Today, we are already critically reliant on space for tools
of communications, observations of climate, and tons of essential
science. In the near future it will be the abundant resources
that draw us further into space. But, in the long run, as risks of
pandemics, killer asteroids, and geopolitical turmoil threaten
our existence, at most we have a few billion years before the sun
literally expands and swallows the Earth. So, humanity *must*
become a multi-planet species, or humanity will become extinct
in the universe.

A CONVERSATION WITH LAETITIA GARRIOTT DE CAYEUX

What first sparked your passion for space?

My interest in space spawned from my passion for advancing humanity, which takes us right back to space in oh so many ways, from the more practical side such as internet broadband (which was born out of NASA), to satellite imagery capable of identifying leaks in pipelines, locating underground water sources that can provide sufficient supplies for entire nations, or enabling ubiquitous connectivity, all the way to the eventual survival of our species. As Bill Nye once pointed out, while the dinosaurs didn't have a space program, we do!

As a non-engineer, you've had an incredible impact in the space industry, including leading a beamed-propulsion technology company. How did you leverage your unique background to create outsized impact?

When Dmitriy Starson shared with me his ideas for a beamed-energy space launch, I immediately recognized that if this could be done, it would be world changing. After conducting a careful review that included independent subject-matter experts, I quickly became convinced that we finally had a shot at making beamed-propulsion technology a reality and I was all ready to go. Dmitriy is a world-class scientist and became a great partner.

While he and the other engineers obviously managed the technical side of things, successful completion of a world-changing idea like this takes more than just a strong engineering team. I helped wrap a real company and plan around their efforts. Other than grit and hard work, what fundamentally allowed us to create outsized impact boiled down to:

1. **Mitigating tail risks.** Keeping a tight rein on things that can kill you, such as a lack of funding. The question you generally want to be asking is how quickly you can get to revenue, then work backward from there. In our case, because there were no ancillary uses for the technology that could provide interim revenues until we build out the full space system, we asked early on what our next fundable milestone should be and what it would take for us to reach this next fundable milestone. This question led us to another question: who could we partner with to cut time to funding and market? The faster you reach that point, the less likely you are to run out of funding on the way. In our case, the ideal partners were NASA and DARPA. It's rare to create outsized impact without partners. Identifying and nurturing these relationships early on helps. Fundamentally, and if I generalize beyond our work at Escape Dynamics to the broader aerospace industry, creating outsized impact is much less about taking risks than it is about mitigating risks that run the gamut from funding to supply chain issues and who will be capable of delivering you that critical part (or not) all the way to faulty analysis such as your dry mass being wrong because you've not accounted for secondary structures and now this rocket ain't gonna fly!

2. **Resource Management.** It is common for well-meaning engineers without proper guidance to focus on the wrong tasks,

for example focusing on a subsystem where they think they can save 50–80 percent of that subsystem's cost, even if that subsystem's overall cost represents a tiny fraction of the overall system cost and the chance of securing that saving is relatively low. Conversely, they could focus on more modest savings on higher-cost items that are both low-hanging fruits and which can drive the total system cost down meaningfully. Good managers make a giant difference in ensuring talent works toward the mission. It can be a challenge with a start-up where the amount of managerial experience is limited. In our case, implementing OKRs ("Objectives and Key Results," a collaborative goal-setting tool used by teams and individuals to set ambitious goals with measurable results created by Andy Grove at Intel) really helped.

You've already completed some spaceflight training; do you intend to travel to space yourself one day? What do you hope to get out of the experience?

My interest in departing the Earth is minimal. But if you ask my husband, he'd love for the whole family to move to Mars!

You were my earliest and most impactful mentor—did anyone invest similarly in you when you were starting out in your career? What motivates you to nurture early talent?

When I was sixteen and the internet just nascent, one of my cousins, who was running a software company, let me know about a global competition where winners would participate in a major international forum in Tokyo and contribute youth perspectives on the Global Information Society and how it could help address major world issues. That door opened many others.

For example, it was at this forum in Japan that I met Professor Negroponte, the founder of the MIT Media Lab, who years later became one of my sponsors on my path to US citizenship. My first employer, Goldman Sachs, also put a strong emphasis on nurturing early talent, and regularly offered women leadership seminars that I found incredibly helpful early in my career. I think nurturing early talent is important everywhere. It's a privilege to do so!

You've also invested in a number of space companies; what sort of technologies are most exciting to you?

A first broad category is sustainably cheaper and more mission-essential technologies. My early investment in SpaceX fits this category. For many years, we kept hearing in Washington that an appropriate risk strategy was to launch high-value payloads on well-proven launch vehicles until high reliability can be achieved by new, lower-cost options. But the model has flipped. The lower-cost option, SpaceX, is also the one with the highest reliability.

Another category is companies with a unique opportunity to bring about a future that is more inclusive where Earth-based technologies have failed. My early investment in Lynk, which demonstrated in 2019 the first "cell tower in space," fits that category. Lynk has the technology and thus a real shot at providing ubiquitous connectivity for anyone on Earth who owns an unmodified cell phone or can invest in purchasing a refurbished one. Earth-based cell towers cannot be a part of the solution as they are and will remain prohibitively expensive to deploy not just in many parts of the developing world but also here in the US. Space-based cell towers can uniquely close that coverage gap. And closing that coverage gap would change the lives of eighty-eight million Americans living in rural areas. And it would also change the lives of over 2.5 billion people in the world who don't have a mobile, not because the handset is too expensive but because they don't live in connected areas.

What industries do you imagine will become increasingly relevant as space exploration capabilities expand?

Space-based resources (minerals/energy), in-space medical R&D, and in-space construction and manufacturing for working and living in space will all become increasingly relevant. Space governance will also be critical to help drive sustainable growth as our space exploration capabilities expand.

What advice would you give to someone looking to break into the space industry as a non-engineer?

Don't apply for the engineering job! :) More seriously my humble piece of advice, if you want to break into the space industry as a non-engineer, is to ask yourself: What is your best use to the space industry's mission—based on what you love to do and what you are great at? What does the space industry need that they will pay for that you love and are great at? And then I'd also say: spend time with the engineers. Not only will you make each other better, and thus augment your impact as a team, but I bet you'll make friends for life. I did.

As a parent, what are your biggest hopes for the next generation? What worries you?

Whether it's the environment, equality, justice, wealth gaps, the digital divide, or even a risk to the very existence of democracy, my concerns for the next generation have expanded these last few years, though fortunately the long-term trend line is still positive. I truly hope the next generation lives in a more inclusive world with more equitable opportunities and where access abounds. And I hope the American experiment of democracy regains its foothold.

Your family is the epitome of a Space Age family (and an inspiration for my own!)—what advice do you have for parents hoping to nurture their children's curiosity about the universe?

My first piece of wisdom is: let your child lead and follow his or her curiosity; that curiosity is innate in them. It's less about trying to awaken curiosity in your child than it is about not stifling it by not listening and making the mistake of forcing our kids to live in an adult-led world. Let the kid lead you and enjoy the journey with her as the lead! What she may notice first about the universe and what she may want to engage with you on may be that water snake she saw and is still in awe of. And that's not just *a* place to start, that's *the* place to start, because that snake sparked her curiosity.

My second piece of advice is: expose your child to as many things as possible because one of these things will spark their curiosity.

You can't force curiosity in anything in particular, but you can help expand your child's horizon until you see curiosity spark!

Great advice. Finally, what drives our desire to explore space? Why should people care?

The rationalist would argue that, fundamentally, if we want humanity to survive and thrive, exploring space is our only path. The humanist would argue that, fundamentally, humans are curious at heart and just like sailing the Atlantic paved the way for the *Mayflower* and settlers expanding into the new world, our generation is ready to leave the confines of Spaceship Earth.

I am not sure how many of the people doing it care about the "why." I'm not sure how we get the nonbelievers on board short of shutting off all GPS communications to help them realize that they unknowingly already cared so much about space. What I do feel strongly about is that humanity's future on Earth is not about "fight or flight," because when we go to space, it's to benefit humanity at large as well as Earth.

A CONVERSATION WITH
LUCY HAWKING

From the moment I met Stephen Hawking's daughter Lucy, I recognized her as a force of nature. She's creative, clever, and fiercely loyal. For me, Lucy was also a role model—a talented writer, a brilliant science communicator, and a loving mother. She's living proof that you needn't be a scientist to contribute to science; the children's books she cowrote with her father fuel the imaginations of the next generation and inspire them to fall in love with all the mysteries of this magnificent universe. It was a career highlight for me to contribute a short chapter to one of those books, George and the Blue Moon.

What were you like as a young girl? What were your interests and aspirations? Who were your role models?

Apparently, I was an exceptionally gregarious young child! I talked pretty much constantly and loved telling jokes, except I often laughed so much myself no one could understand the punch line. I don't know that I had role models as such—I don't think the whole concept of role models was as prevalent in those times. But I think it's a thoroughly good thing that young people are now encouraged to think about whom they admire and why, and which qualities they want to take up in their own lives. I was very interested in the performing arts—theatre, dance, drama. I even hosted my own chat show aged around seven years old; I interviewed family members and recorded episodes of my show on cassettes!

Your dad was larger than life—what was it like having a scientific luminary as a parent? Did you feel any societal pressure to study science?

My father was a superstar and I miss him terribly. Although it's been two years since he died, I still find it quite painful to think that he is not there anymore in his physical form. Part of me still expects him to just pop up somewhere and for life to go on as it always did. But then I still feel so connected to him through our family, through memory, through his work, and through an appreciation of the magnificence of the universe in which we live. My father would have liked me to study science but in fact, when I went my own way and studied arts subjects, he accepted that I needed to make my own choices in life.

What first sparked your curiosity about space and the universe?

I remember being incredibly small and looking through an absolutely enormous telescope. I think this was at the Institute of Astronomy in Cambridge. The realization that we could so clearly see something so far away blew my mind. It showed me that there were worlds far beyond ours and that something exciting and enormous existed past the envelope of our atmosphere. I probably didn't put it in quite those terms aged three, though.

You've built your own space career as a renowned author and science communicator; what unique strengths and perspectives do you think non-engineers bring to the space industry?

I started my career in science writing (I was already a writer by this stage) with an attempt to put my father's work into a narrative context for young readers. I realized that most people don't have the facility to understand abstract concepts and get baffled and lost quite quickly. I had the idea that if we could find a way to put scientific information into storytelling, to frame it in terms of human experience, that would make it so much easier to comprehend. The arts in general are so brilliant at engagement, at pulling you in emotionally and making you care about the subject matter, and I felt this was the missing element in science communication at that time. No one was trying to make science as compelling and exciting as a really good novel, film, or series, despite the fact that the subject matter—the universe itself—was ultimately fascinating, bizarre, and intense. Perhaps because I came at this as a storyteller and not a scientist, I was freer to take liberties than someone from a more classically scientific background might have been? Although I have never broken the laws of physics for a story, I'd like to add.

You and your dad gave me the greatest gift in inviting me to contribute to your George adventure series. So much of your work centers around teaching science to children and exposing the next generation to the wonders

of the cosmos—what made you passionate about STEM education?

Kellie, we were both such great fans of yours and it was a huge honor to have your brilliant voice in our book! It's been such a joy, working with you, being friends, seeing you and your family flourish! You are such an exceptional human being, not something I say very often. I am a huge believer in the value of education as the greatest investment we can make in the future of our world. If we as educators and adults can instill confidence, encourage talent, grow independence, promote kindness and compassion, develop awareness of our environment, our shared humanity, our responsibility for others, and spark creativity, innovation, and invention through education, then the world has a chance. STEM education is incredibly important as the world is going to need brilliant scientists and technicians to solve some of the really huge challenges that we see already—climate change, pandemics, destruction of species, acidification of the oceans, spreading deserts, extreme weather, and growing inequality to name but a few—but the world needs people with many different types of ability. So, I'm promoting STEAM education (STEM plus the arts) as the way forward!

As a parent, what's your biggest hope for the next generation? What's your biggest hope for the future of space exploration?

My biggest hope for the next generation is that they get as many opportunities as we did—and that they use them better! My big hope for the future of space exploration is that the human race gets out to space again and goes further than ever before. I'd like to see an astronaut walk on Mars, and yes, I would be excited if that space traveler were a woman.

Do you believe we're alone in the universe?

Absolutely not! I am entirely sure there is life out there somewhere. Whether it is intelligent life and whether we ever meet it are more complex questions.

You've been a mentor to me, and it has been so meaningful to watch another woman (and mother!) navigate such a successful career in science communication. What are your thoughts on the current state of representation in the space industry? What does it mean to you?

I think it is kind of astonishing that it has taken so long for diversity to find its rightful place in the space industry. It's part of the legacy of the military history of space, that it was seen as the domain of conquering heroes, the square-jawed Navy test fighter pilots who had the "right stuff" to go into space. Interestingly, the Soviet Union sent a woman, Valentina Tereshkova, into space very early on in the Space Race, but even that was a heavily politicized statement about the superiority of the communist regime rather than a genuine desire to promote equality. The second woman in space, Svetlana Savitskaya, had quite a ropey time of it in space, with cosmonauts presenting her with flowers and an apron when she arrived on the Russian space station. The whole question of gender and space was treated with something akin to fear and trepidation until female astronauts started traveling regularly into space, and of course, it was obvious that these random "fears" were totally unfounded. There

NOT NECESSARILY ROCKET SCIENCE

is still a long way to go before our marvelously diverse human race on Earth is properly reflected in the astronauts who travel to space, but there definitely has been a giant leap. Thankfully.

Why does human space exploration matter? What's at the heart of the human quest and passion for the cosmos?

To ask why human exploration matters is like asking why do humans matter. Because if you only sent robots out into space, it would be akin to saying that, in the grand scheme of things, human beings don't count for much. If we look at the most exciting, the grandest, the most challenging, evocative, and society-changing exploration that exists, that of a journey to the Moon, Mars, and beyond, and say we only send our robot friends, then we've just opted the human race out of the future. To explore is to go boldly into the future and if we don't trust actual human beings to go out and fashion that future for the rest of us, then we might as well give up and hand our whole civilization over to robots to run. Space exploration matters for so many reasons: philosophical, technical, scientific, societal, and those big reasons need human beings who can rise to the challenge of fulfilling them. At least, that's what I think.

There are many brilliant cosmologists, but very few can capture the imaginations of the general public the way your dad was able to; what do you think made him such an effective science communicator?

It's something to do with his astonishing intelligence, his vast and capacious memory, and his ability to distill everything down to the absolute bare minimum without losing any impact or meaning. He had great deadpan timing too. We used to speculate about alternative careers for him and stand-up comedian would have definitely been one. There was an X factor about him—when he spoke, people listened. It was quite extraordinary.

I'll always remember the hauntingly beautiful celebration of Professor Hawking's life and achievements at Westminster Abbey after his passing. There wasn't a dry eye as his ashes were interred in the nave between the graves of Sir Isaac Newton and Charles Darwin, a new memorial stone lowered to its rightful place among the scientific luminaries who have contributed so much to our never-ending quest to understand the world around us. When my daughter is a bit older, I plan to bring her back with me to visit. I can't wait to share with her the memorial of a great man with whom her existence ever so briefly overlapped.

A CONVERSATION WITH MICHAEL LÓPEZ-ALEGRÍA

What were you like as a child and what were your earliest career ambitions? Were you always interested in space or did that develop later?

My mom used to work for NASA and would bring home these pamphlets called "NASA Facts" about their early satellites and of course human spaceflight programs, so I was a space nerd before it was cool to say that. I remember vividly the Eagle's descent to the lunar surface, as I was with my family at the beach, huddled around a transistor radio. But in middle school and high school my gaze wandered. I did not become a Navy pilot with an eye toward NASA, but later at age twenty-five, when I became interested in becoming a test pilot and read about so many graduates of the US Naval Test Pilot School that had gone on to become astronauts, the childhood dream was reborn.

What motivated you to apply to be a NASA astronaut? What was your experience like during the selection process, and what was your reaction when you found out you had been selected?

I was lucky enough to have been asked to do an interview and medical exams (basically the final round) the first time I applied for the 1990 class, but did not prepare well for the interview (and arrived for it late—don't ask!) The second time was smoother sailing. Getting the call was surreal—like spaceflight itself. When you pinch yourself, it hurts. But some part of you still can't believe it.

You've spent more time in space than nearly any other human being in history; did the novelty ever wear off? What surprised you about living in space? What did you miss on Earth?

Perhaps what surprised me the most is that the novelty did *not* wear off. I always thought of myself as a sprinter who would do poorly in a marathon, but I ended up enjoying it that much more. Besides my loved ones, I missed the simple things—the smell of fresh-cut grass or summer rain on asphalt, the feeling of wind in your face, the taste of wine at dinner.

Did you experience a cognitive shift after visiting space yourself? How has your spaceflight experience affected your outlook on life?

I most definitely experienced the cognitive shift known as the overview effect. It's rather subtle, but undeniable. One comes back more aware of how connected all of us that sail on the Spaceship Earth are, and of how we must strive to do better to get along. And that we must take care of our precious planet.

After a successful NASA career, you became a powerful advocate for the commercial spaceflight industry—what motivated you to champion the growth of a private sector?

My attitude started changing from that of a steely-eyed test pilot to one of a space democratization evangelist with my experience training and flying with Anousheh Ansari for *Soyuz TMA-9* in

2006. She had a very different—and more global—perspective of human spaceflight than anything to which I'd been exposed. It turns out human space exploration should not be only about government space programs; the more people that experience it, the better a place our world becomes.

You took a chance on me early in my career; what unique strengths and perspectives do you believe non-engineers can bring to the space industry?

Plenty. I've become increasingly aware of the richness of diversity—between left- and right-half brain thinkers. Engineers are inherently risk averse; great accomplishments are often the result of bold initiatives.

What future space industry milestones, either near-term or far, are you most looking forward to?

Human spaceflight may someday become routine, but I don't know if I'll be around to witness that. I do think it will become increasingly commonplace and I look forward to when we have increased the number of space travelers by an order of magnitude. I'll have to live a long time, but I'm optimistic.

If you headed to space again, what would be your ideal mission?

I'd love to walk on the Moon, but I honestly don't see that happening. What's the art of the possible? Shooting a movie with Tom Cruise aboard the ISS? (Just kidding...)

As a parent, what are your hopes for how the next generation benefits from space?

I am somewhat ashamed of the mess my generation will leave for the next. The serious problems seem too many to count, and the trend of nationalism and me-centrism around the world does not bode well for finding solutions. But I am very impressed with the knowledge that the information age has brought to young people, and even more by their corresponding curiosity and thirst for it. That, combined with a healthy number of them experiencing the overview effect, gives me optimism that they will right the ship.

Why should people care about space exploration? Why is opening up access to space so important?

The same tried and true answers will always apply—spin-off benefits, international cooperation, pushing the boundaries of technology, etc. But the fundamental reason we explore is because humans always want to know what's around the next bend; to answer the essential questions about life; to go where no (wo)man has gone...it's just who we are.

AUTHOR'S NOTE

Summer 2020

It's hard to understate the rapid rate of progress in the space industry. The last ten years have changed the future of space exploration, but even in the last approximately six months, the landscape has evolved significantly:

- In December 2019, the Air Force Space Command was independently established as the United States Space Force, the sixth and youngest branch of the US Armed Forces, with a mission to organize, train, and equip space forces in order to protect US and allied interests in space and to provide space capabilities to the joint force.

- In March 2020, after a nationwide naming contest, NASA JPL introduced the world to Perseverance, nicknamed "Percy," the newest Mars rover launching to the red planet to seek signs of ancient life and collect rock and soil samples for potential return to Earth.

- In April 2020, NASA selected my underdog alma mater Masten Space Systems to deliver scientific cargo to the Moon's South Pole in advance of America's return to the lunar surface. (Artemis, twin sister of Apollo, is the name of NASA's new lunar program that will carry the first woman and the next man to the Moon in an effort to establish an ongoing presence beyond low Earth orbit.)

- In May 2020, a decade of work with NASA's Commercial Crew Program culminated in SpaceX restoring US access to the International Space Station aboard American spacecraft. The *Crew Dragon Demo-2* launch was SpaceX's first ever crewed flight and NASA's first mission to launch astronauts into space from US soil since the space shuttle program ended in 2011.

- And so on...

All this amidst a backdrop of a global pandemic and an overdue confrontation with racial inequality, underscoring both the preciousness of a single human life and the precariousness of our long-term survival as a species. The contrast and emotional dissonance between exhilarating space achievements and devastating Earthly happenings is only going to grow as this decade unfolds. These are the highs and lows of life in the Space Age and a stark reminder of why we continue to push the boundaries of our knowledge and capability. Now, more than ever before, we are attempting to survive our time so that we may live in the future. Echoing our earliest emissaries to the stars, we cling to our hope and our determination, and our good will in a vast and awesome universe. Ad Astra!

—KELLIE

RECOMMENDATIONS

If you're looking to launch yourself into the space community, here are a few places to start!

Annual Conferences and Ongoing Events

1. NASA Socials
2. The NewSpace Conference
3. The International Space Development Conference (ISDC)
4. The Space Symposium
5. The Small Satellite Conference
6. Humans to Mars (H2M) Summit
7. The International Astronautical Congress (IAC)
8. Spacefest
9. Yuri's Night
10. The Goddard Memorial Dinner (a.k.a. "Space Prom")
11. FAA Commercial Space Transportation Conference

Organizations

1. Students for the Exploration and Development of Space (SEDS)
2. The Space Frontier Foundation
3. The Planetary Society
4. The Mars Society
5. Explore Mars
6. The Commercial Spaceflight Federation

International Educational Programs

1. Space Camp USA
2. International Institute of Astronautical Sciences (IIAS) and Project PoSSUM
3. International Space University (ISU)

Scholarships and Fellowships

1. The Brooke Owens Fellowship
2. The Matthew Isakowitz Fellowship
3. Project PoSSUM's "PoSSUM 13" and "Out Astronaut" competitions

TOP FIVE FAQS

(AS UPVOTED ON SOCIAL MEDIA)

1. Do you believe intelligent life exists beyond Earth?

I want to believe! Here are a few factors I love to consider when contemplating the possibility: Assuming trillions of galaxies, let's estimate one septillion stars in the observable universe. A large number of those stars likely have planetary systems of their own, and a good deal of those planetary systems may contain planets in the "habitable zone," with the right conditions for liquid water to form on their surface. And many of those "habitable zone" planets could be Earth-like in size and mass. And...

2. What's the significance of your daughter's name?

My daughter's name is Delta Victoria. The "Delta-V" is a nerdy reference to the flight dynamics of a spacecraft, symbolized as Δv and more literally translating to "Change in velocity." The name was also apt; her arrival certainly changed the velocity of our family's life.

3. How do you suggest we get more women in STEM?

I don't think it's as simple as increasing exposure (although that's certainly important!). We need to address both the underlying pipeline problems and some serious root obstacles, like the wage gap, reproductive rights, and paid parental leave. It can be frustrating to watch powerful men in America hit the peak of their careers well into their seventies (looking at you, politicians!), while women tend to get cut off at the knees around thirty. The future we want requires making adjustments in our present, and I've learned firsthand that these issues go hand in hand.

4. What's your favorite space fact?

Theoretical physicist Brian Green remarked that if you hold your thumb at arm's length against the night sky, you'll cover more than ten million galaxies in the observable universe. I knew it academically, but the scale of the visual blew my mind. Space is big!

5. What's your favorite sci-fi film?

It's impossible for me to choose one, so I'll share my top five: *Gattaca*, *Sunshine*, *Solaris*, *Moon*, and *Event Horizon*. Enjoy!

ACKNOWLEDGEMENTS

Like Ann Druyan and the *Voyager* team constructing the golden record, I am sitting around a kitchen table agonizing over what to include and what to leave off. As I prepare to send my little book out into the world, I'm overwhelmed by a sentimental desire to list here everyone who has ever been deserving of thanks in my life—so I can only imagine the pressures of curating a cultural Noah's Ark! But if President Carter could keep his galactic well-wishes to a short paragraph, I can aim to keep my thanks succinct.

I'm grateful first to my parents (known now as Grandma and Pop Pop) and my large, loving, and globally dispersed extended family for encouraging me to reach for the stars and supporting me in every way as I continued to take that direction literally. I'm grateful to my husband Steven for his sense of humor and his constant encouragement to explore everything that this universe has to offer me. And I'm grateful to our daughter Delta, who always reminds me that home is the best place to be.

I'm forever in awe of my extraordinary colleagues, past and present, at the Commercial Spaceflight Federation, each of whom has played an important role in ushering in a new era of commercial human spaceflight. Special thanks in particular are owed to Mike L-A's successor Eric Stallmer for his endless support of my personal and professional goals and for always keeping the porch light on for me at CSF, no matter how far off course my career journey took me.

As mentioned throughout the book, I'm indebted to a number of mentors and managers who not only provided career direction but who also held doors open and encouraged me to march through them with confidence. I'm especially grateful to Laetitia and Richard Garriott de Cayeux, Mike L-A, Sean Mahoney, and Jubal Stout for making bets on me when it counted.

Speaking of bets, enormous thanks are also owed to my publishers at Mango, without whom this book wouldn't exist. Thanks in particular to my editor Hugo Villabona for believing in this book when it was just an idea and for guiding me through the process of bringing it to life.

I'm endlessly grateful to the hundreds of thousands of people who have followed my journey on social media and helped create a community of passionate space advocates—you've enriched my life over the course of a decade and it's because of you that we're able to take this conversation offline and onto book shelves.

Finally, I want to thank each and every one of you, dear readers, for picking up this book and allowing me the honor of welcoming you to the Space Age. Buckle up, though—no one promised a smooth ride.

ABOUT THE AUTHOR

@KellieGerardi is committed to making the most of life in the Space Age. As an aerospace & defense professional and popular science communicator, Kellie has flown multiple microgravity research campaigns as a citizen scientist and space-suited human test subject. She is a scientist-astronaut candidate with Project PoSSUM's suborbital research program and leads special projects for the Commercial Spaceflight Federation. Kellie's work to promote citizen-science and inspire women in STEM has been featured across a broad range of media and has attracted hundreds of thousands of fans on social media. Kellie also serves on the Defense Council for the Truman National Security Project and is an active member of The Explorers Club, whose esteemed flag she carried during a crew rotation at the Mars Desert Research Station. Kellie lives in the Washington, DC, area with her husband Steven and their daughter, Delta V. *Not Necessarily Rocket Science* is her first book.

"GET IN LOSER, WE'RE GOING TO SPACE."

Mango Publishing, established in 2014, publishes an eclectic list of books by diverse authors—both new and established voices—on topics ranging from business, personal growth, women's empowerment, LGBTQ studies, health, and spirituality to history, popular culture, time management, decluttering, lifestyle, mental wellness, aging, and sustainable living. We were recently named 2019 *and* 2020's #1 fastest growing independent publisher by *Publishers Weekly.* Our success is driven by our main goal, which is to publish high quality books that will entertain readers as well as make a positive difference in their lives.

Our readers are our most important resource; we value your input, suggestions, and ideas. We'd love to hear from you—after all, we are publishing books for you!

Please stay in touch with us and follow us at:

Facebook: Mango Publishing
Twitter: @MangoPublishing
Instagram: @MangoPublishing
LinkedIn: Mango Publishing
Pinterest: Mango Publishing
Newsletter: mangopublishinggroup.com/newsletter

Join us on Mango's journey to reinvent publishing, one book at a time.